The Development of Non-Financial Reporting

Valentina Minutiello
Editor

The Development of Non-Financial Reporting

The Role of Sustainability Reporting
and Integrated Reporting in Corporate Strategy

Editor
Valentina Minutiello
Carlo Cattaneo University, LIUC
Castellanza, Italy

ISBN 978-3-031-83180-5   ISBN 978-3-031-83181-2 (eBook)
https://doi.org/10.1007/978-3-031-83181-2

© The Editor(s) (if applicable) and The Author(s), under exclusive license to Springer Nature Switzerland AG 2025

This work is subject to copyright. All rights are solely and exclusively licensed by the Publisher, whether the whole or part of the material is concerned, specifically the rights of translation, reprinting, reuse of illustrations, recitation, broadcasting, reproduction on microfilms or in any other physical way, and transmission or information storage and retrieval, electronic adaptation, computer software, or by similar or dissimilar methodology now known or hereafter developed.
The use of general descriptive names, registered names, trademarks, service marks, etc. in this publication does not imply, even in the absence of a specific statement, that such names are exempt from the relevant protective laws and regulations and therefore free for general use.
The publisher, the authors and the editors are safe to assume that the advice and information in this book are believed to be true and accurate at the date of publication. Neither the publisher nor the authors or the editors give a warranty, expressed or implied, with respect to the material contained herein or for any errors or omissions that may have been made. The publisher remains neutral with regard to jurisdictional claims in published maps and institutional affiliations.

This Palgrave Macmillan imprint is published by the registered company Springer Nature Switzerland AG.
The registered company address is: Gewerbestrasse 11, 6330 Cham, Switzerland

If disposing of this product, please recycle the paper.

A Short Premise

The volume addresses the topic of non-financial reporting, focusing on the two main types of reporting: Sustainability Reporting (SR) and Integrated Reporting (IR).

Reporting tools are now the subject of great attention from the academic world, due to the growing international regulatory pressure on the communication of non-financial information. In the current context, in fact, it is now widely believed that it is essential to convert the business world in a sustainable way.

This solution responds to the widely held concept of "resilience", which implies continuous readjustment to meet the needs of an external environment in continuous evolution and, consequently, to endure over time. This concept is in harmony with that of sustainable development, which in addition implies the ability to satisfy the needs of current generations without compromising those of future generations.

Non-financial communication, therefore, bears witness to this commitment undertaken by companies, describing the actions undertaken in favor of sustainability and the positive and negative impacts of the various businesses.

To date, it is believed that there is still a long way to go to achieve clear and complete corporate communication on these. There are numerous frameworks that can be adopted and discrepancies in the approaches and types of documents produced, aspects that often undermine the effectiveness and quality of non-financial reporting.

Over the years, the academic world has been trying to provide its contribution to the development of the topic, analyzing the structure and quality of the different types of non-financial reports produced and identifying the main factors that can help companies produce high-quality reporting.

The volume in question is intended as an overview (not exhaustive) of the research on the topic, with the aim of highlighting the diversity of approaches used by researchers as well as the multiplicity of facets that concern non-financial reporting. For this reason, it adopts an empirical approach and does not provide a theoretical treatment of the topic, but rather collects some empirical contributions conducted by different authors. The aim is to provide an overview of the most current and interesting research topics. Each chapter is dedicated to a different theme and will report, first, a literature review of the topic and, second, an empirical analysis, thus allowing to appreciate also the application of the different methodologies used.

The chapters will address the following subjects. Chapter 2 is dedicated to the development of a Systematic Literature Network Analysis (SLNA) on Intellectual Capital Disclosure, since the topic of intangibles is still today the subject of great debate with reference to their measurement and their relationship with sustainability. Chapter 3 takes into account a specific type of disclosure, that on the circular economy, and conducts a study in the European context. Finally, the last chapter evaluates the impact of negative environmental social governance (ESG) news on companies' ESG disclosure.

Although the chapters may seem to be little connected, they address the main macro-themes related to non-financial disclosure and attributable to the following categories:

1. the factors that influence the quality of non-financial reports;
2. the different types of non-financial disclosure (such as circular economy disclosure or intellectual capital disclosure);
3. the motivations behind the adoption of non-financial communication explained, for example, according to the legitimacy theory as the need for companies to improve or restore their reputation on the market.

For each category, the volume provides examples of studies and starts the discussion for further necessary insights.

Acknowledgments

I sincerely thank my co-authors for their support and collaboration over the years. I also thank the entire scientific community for the lively exchange of opinions and knowledge.

Contents

1. The Evolution and New Trends of the Intellectual Capital Disclosure: A Systematic Literature Network Analysis   1
   Valentina Minutiello, Anna Lucia Missaglia, and Patrizia Tettamanzi

2. The Circular Economy Disclosure in EU Setting   45
   Isabel-María García-Sánchez and Saudi-Yulieth Enciso-Alfaro

3. Is It Just a Matter of Reputation? A Study on the Impact of ESG Controversies on Corporate Disclosure   67
   Valentina Minutiello and Patrizia Tettamanzi

Conclusions   93

Index   97

Notes on Contributors

Saudi Yulieth Enciso Alfaro is affiliated with the research group GobSal (Corporate Governance, Sustainability and Information Systems), led by Isabel-María García-Sánchez of the University of Salamanca. She is currently engaged in a research programme for young researchers at the University of Salamanca. Furthermore, she is interested in pursuing research in the following areas: circular economy, sustainability, leadership, and gender. Her contributions to academic publications include articles, book chapters, and conference presentations.

Isabel-María García-Sánchez is Full Professor of Accounting at the Universidad de Salamanca (Spain). She has Ph.D. in Economics and Business Administration from the Universidad de Salamanca (Spain). She writes for interdisciplinary indexed journals. Her research interests include sustainability, corporate governance as well as accounting.

Valentina Minutiello (Ph.D.) is a Research Fellow in Accounting at Cattaneo University (LIUC) and a Certified Chartered Accountant. At the same University, she is an assistant professor for the courses Management and Principles of Accounting, Accountability and Sustainable Accounting, Accounting and Financial Reporting, and Impact Accounting and ESG Reporting. Since she entered into the academic world as a Ph.D. student in Management, Finance and Accounting, she has shown interest in the following research topics: the adoption of Sustainability Reporting and/or Integrated Reporting and the identification of factors that can increase their quality, the evolution of Accounting Education, corporate governance issues and their implications for corporate reporting, the peculiarities of family businesses and SMEs concerning sustainability, the Intellectual Capital Disclosure and the Diversity, Equity and Inclusion practices. As a demonstration of her commitment, there are several scientific articles and books in both national and international contexts and journals.

Anna Lucia Missaglia (Ph.D.) is a Research Fellow at Bocconi University. Her research interests are in the area of Management Accounting and Controls, and revolve around Sustainability, Diversity Equity and Inclusion and other social issues. She is author of scientific articles and books within the national and international context. Anna is Teaching Assistant of Planning and Control (B.Sc.) and Corporate Planning (M.Sc.) at LIUC—Cattaneo University.

Patrizia Tettamanzi (Ph.D.) is Full Professor of Financial Accounting and Sustainability Management at Cattaneo University (LIUC) and Senior Lecturer in Financial Reporting and Analysis at Bocconi University. Her main research areas are international financial and nonfinancial reporting, sustainability, accounting education and auditing. She is the author of many publications in international journals. She is a certified chartered accountant and an auditor.

Abbreviations

BNA	Bibliographic Network Analysis
BoD	Board of Directors
CE	Circular Economy
CEO	Chief Executive Officer
CNA	Citation Network Analysis
CSR	Corporate Social Responsibility
CSRD	Corporate Sustainability Reporting Directive
ESG	Environmental Social Governance
ESRS	European Sustainability Reporting Standards
EU	European Union
FE	Fixed Effect
GCS	Global Citation Score
GDP	Gross Domestic Product
GLCS	Global Local Citation Score
GRI	Global Reporting Initiative
IC	Intellectual Capital
ICD	Intellectual Capital Disclosure
IIRC	International Integrated Reporting Council
IFRS	International Financial Reporting Standards
IPO	Initial Public Offering
IR	Integrated Reporting
IRS	Integrated Reporting Scoreboard
KNA	Keywords Network Analysis
NFR	Non-Financial Reporting

NGO	Non-governmental organization
OLS	Ordinary Least Squares
RE	Random effect
RBV	Resourced-Based View
ROA	Return on Assets
SLNA	Systematic Literature Network Analysis
SLR	Systematic Literature Review
VRF	Value Reporting Foundation
WGI	Worldwide Governance Indicators

List of Figures

Fig. 1.1	The largest component of the ICD literature network. Source: Processing by Pajek	9
Fig. 1.2	Distribution of scientific articles published during the years 1999–2022. Source: Personal reworking	9
Fig. 1.3	The Main Path of the biggest connected component of the citation network. Source: Processing by Pajek	10
Fig. 1.4	Resulting clusters of the co-occurrence network of author keywords analysis, based on algorithm of the software VOSviewer. The colours red, green, blue, yellow and purple represent, respectively, cluster number 1, 2, 3, 4 and 5. Source: processing by VOSviewer	26
Fig. 1.5	Co-occurrence network of author keywords according to the year of publication. Source: Processing by VOSviewer	27
Fig. 1.6	Burst detection algorithm applied to normalized author keywords from 2001 to 2022. Source: Processing by Sci2 software	29
Fig. 2.1	Dynamic and geographic evolution of ScoreIRUCE: resources use and circular economy information disclosure by year and country. (a) Average of the interannual evolution of information disclosure. (b) Average of information disclosure by country. Source: Own elaboration	56
Fig. 2.2	Dynamic and geographic evolution of sub-scores of resources use and circular economy information disclosure by year and country. (a) Average of information disclosure evolution by year. (b) Average of information disclosure by country. Source: Own elaboration	58

Fig. 2.3	Disclosure sub-scores for industry. Source: Own elaboration	59
Fig. 2.4	Disclosure by industries risk levels on the environment. (**a**) Percentage of companies on the high, medium and lower levels. (**b**) Average ScoreIRUCE by risk levels on the environment	60

List of Tables

Table 1.1	Main Path papers, with evidence of the title, author, journal, number of citations and year of each publication	12
Table 1.2	Global Citation Score (GCS) of the ten most cited papers, together with title, authors, journals and years of publication	19
Table 1.3	Global-Local Citation Score (GLCS) of the ten most cited papers published in 2022, 2021, 2020 and 2019, together with title, authors and journals	22
Table 1.4	Author keywords in each of the three identified clusters	27
Table 1.5	Research directions in the ICD field from a methodological point of view	32
Table 2.1	ScoreIRUCE: composition	52
Table 3.1	Sample	75
Table 3.2	Descriptive statistics	78
Table 3.3	Descriptive statistics by year	78
Table 3.4	Pearson correlation coefficients (*** $p < 0.01$, ** $p < 0.05$, * $p < 0.1$)	80
Table 3.5	Panel fixed effects regression	81

CHAPTER 1

The Evolution and New Trends of the Intellectual Capital Disclosure: A Systematic Literature Network Analysis

Valentina Minutiello, Anna Lucia Missaglia, and Patrizia Tettamanzi

Abstract The purpose of this chapter is to extract the backbones of the research field on Intellectual Capital Disclosure (ICD) and to identify new directions for future research, by answering two different research questions:

RQ1: What are the main themes that have been developed within ICD research?
RQ2: What is the future of ICD research?

V. Minutiello (✉) • P. Tettamanzi
Carlo Cattaneo University, LIUC, Castellanza, Italy
e-mail: vminutiello@liuc.it; ptettamanzi@liuc.it

A. L. Missaglia
School of Economics and Management, Carlo Cattaneo University—LIUC, Castellanza, Italy

Bocconi University, Milan, Italy
e-mail: amissaglia@liuc.it

© The Author(s), under exclusive license to Springer Nature Switzerland AG 2025
V. Minutiello (ed.), *The Development of Non-Financial Reporting*,
https://doi.org/10.1007/978-3-031-83181-2_1

An innovative literature review method was applied, which combines a systematic literature review and bibliographic network analysis: the "Systematic Literature Network Analysis" (SLNA). Moreover, the SLNA was integrated with other tools, such as the Global Citation Score analysis and the keyword analysis.

The analysis suggests the presence of two directions in the development of the field: the first is related to the identification of the main determinants of the ICD quality and diffusion, and the second regards the design of a shared theoretical framework to support further investigations on this topic.

The main limitations concern the use of only one database (Scopus) for extracting the data and the presence of subjectivity problems. However, these limitations are partially solved through different tools to perform the analysis.

This chapter contributes to the literature by providing a systematic view of the existing debate and new unexplored streams of research on Intellectual Capital Disclosure. Besides, it highlights the quality of a new methodology (called SLNA) to conduct a dynamic analysis in this research field.

Keywords Intellectual capital disclosure • Quality • Determinants • Literature review • Systematic literature network analysis

1.1 Introduction

Today there is an increasing demand for financial and non-financial information (Eccles & Mavrinac, 1995; Campanella et al., 2013; Barile et al., 2015) that aims to show how companies create value over time (Lev, 2001) and how they manage their knowledge (Bukh et al., 2001). In this context, the need to know more about companies' intangible assets has emerged, given their crucial role in the value creation process and their contribution in creating and maintaining a competitive advantage over time (Skoog, 2003; Hsu & Fang, 2009; Li et al., 2010; Kong, 2010; Lin & Huang, 2011; Martín et al., 2011; Roulstone, 2011). Moreover, companies have recognized that intangible assets are a driver of their performance (Stewart, 1997; Sveiby, 1997; Lev, 2001; Greenhalgh & Longland, 2005; Hyvonen & Tuominen, 2006; Abeysekera, 2006; Sharma et al., 2007; Sällebrant et al., 2007; Yi & Davey, 2010).

Intangible assets can be defined as non-physical resources that produce positive returns over a long-term period (Li et al., 2010; Abhayawansa & Guthrie, 2014; Singh & Narwal, 2015; Castilla-Polo & Gallardo-Vázquez, 2016). Their aggregation is called Intellectual Capital (IC) (Castilla-Polo & Ruiz-Rodríguez, 2017). It is possible to find many different definitions of Intellectual Capital in the literature (Sveiby, 1997; Stewart, 1997; Schneider & Samkin, 2008; Kwee, 2008; Yi & Davey, 2010; Dumay, 2014; Chiucchi & Dumay, 2015; An et al., 2015; Dumay, 2016). Traditionally, it is divided into three categories: Human Capital, Organizational Capital and Relational Capital (Sveiby, 1997). Human Capital refers to individual knowledge and human resources (e.g. competencies, skills, education, values and experiences) (Guthrie et al., 1999; Pablos, 2002). Structural or Organizational Capital involves the organizational structure, processes, procedures, systems and culture (Guthrie et al., 1999; Pablos, 2002; Wong & Gardner, 2005). Relational Capital is related to the external relationships of the organization with suppliers, customers, business partners and the other stakeholders (Wong & Gardner, 2005).

Coherently with these premises, Intellectual Capital Disclosure (ICD) has gained more and more attention from researchers. Previous contributions suggested that there is an increasing trend on the part of firms to provide additional information on intangibles (Williams, 2001; Abeysekera & Guthrie, 2003; Bozzolan et al., 2003; Vandemaele et al., 2005; Gray et al., 2004; Burgman & Roos, 2006, 2007). Moreover, ICD became more relevant in accounting literature because of its inclusion into a new reporting tool: Integrated Reporting (IR) (Integrated Reporting Framework, 2021). Indeed, Intellectual Capital is embedded in the six capitals of the IR fundamental concepts (IRF, 2021) that are stocks of value used by companies in the process of value creation.

However, the literature on ICD is still fragmented, with contradictory findings and the absence of consistent pieces of evidence on the benefits of this kind of disclosure (Dumay & Cai, 2014, p. 279). Consequently, some authors have highlighted the need to study this topic more in depth in order to systematize the current state of the art (Edvinsson, 2013, p. 163). In addition, due to the large number of studies in the accounting field (both qualitative and quantitative), there is a need for properly classifying and quantifying previous contributions to identify some possible new trends for research (Colicchia & Strozzi, 2012; Dumay, 2014; Massaro et al., 2016). Specifically, Afeltra et al. (2022) called for more literature contributions on the topics of non-financial disclosure quality and

Integrated Reporting, which include the theme of ICD, being intellectual capital a relevant content of Integrated Reporting. The authors applied and suggested an innovative and complex research methodology, according to Strozzi et al. (2017), which has the strength of analysing an individual field from different perspectives, using both bibliometric and citation-based tools.

This chapter, hence, has answered the call by studying ICD as a network with a citation-based approach and considering citations as a proxy of studies' relevance (Strozzi et al., 2017). A Systematic Literature Network Analysis (SLNA) (Strozzi et al., 2017) could be useful to understand the actual focus of Intellectual Capital Disclosure research and directions for future research. The SLNA was integrated with other tools such as the Global Citation Score (GCS) analysis and the keyword analysis (Strozzi et al., 2017). Afeltra et al. (2022) proved the accuracy and effectiveness of the methodology to study Corporate Social Reporting; hence, being ICD an adjacent topic, a positive outcome could be expected from the application, enabling to study both the current state of art and the evolution of the field over time.

To this end and consistently with Cuozzo et al. (2017) and Waltman et al. (2010), this chapter tries to answer the following research questions:

RQ1: What are the main themes that have been developed within ICD research?
RQ2: What is the future of ICD research?

The chapter is structured as follows. Section 1.2 presents the description of the bibliographical material, the methodology and the discussion of the results related to the SLNA's first and second phases, as well as the Global Citation Score analysis and the keyword analysis. Finally, Sect. 1.3 shows main research findings and some suggestions for further research.

1.2 Methodology and Application

Material and Methods

The SLNA is conducted on the data collected from the Scopus database. Various scholarly citation databases are available (some examples are Web of Science, Google Scholar and Scopus), but the best solution is to refer to Scopus. One reason is that Google Scholar, even if it is the largest database and is completely free, also includes other sources (such as administrative notes, library tours, student handbooks) that are not traditional scholarly material (Noruzi, 2005) while Web of Science is smaller with respect to Scopus. The coverage of Scopus is almost 60% larger than that of Web of Science (Zhao & Strotmann, 2015). Finally, Scopus is also the most commonly used database (Strozzi et al., 2017).

The process of SLNA is divided into different phases. During the first phase, a systematic literature review (SLR) is performed using Scopus database to define the scope of the analysis and its boundaries, to conduct the locating studies of "keywords, time, type of documents, language" and to obtain the study selection and evaluation, which are useful to isolate the most relevant works (Rashman et al., 2009; Carter & Easton, 2011). All these steps will be explained better in the following paragraphs. This technique is more objective than other traditional literature reviews, thanks to its principles (transparency, inclusivity, explanatory and heuristic nature) (Denyer & Tranfield, 2009; Kim et al., 2018, p. 1034) and also to the fact that it allows us to obtain a first selection of papers to be included in the following analysis.

The second phase is a bibliographic network analysis (BNA). The inputs of this phase are the papers selected during the first phase. There are three steps to follow here. The first step is the citation network analysis (CNA), which implies an aggregation of papers into communities based on their contents and the identification of those that contribute more to the scientific field. It is based on the assumption that researchers in the same field tend to cite each other. The most relevant citations can be divided into different paths and from them it is possible to identify the Main Path: it is the "backbone of the research tradition" (Lucio-Arias & Leydesdorff, 2008; De Nooy et al., 2011) as it is useful in selecting papers that are the main reference points for recent studies (Strozzi et al., 2017). Being the study of Main Path, a citation analysis based on social networks, two main software are involved in this step, i.e. VOSviewer to extract the biggest connected component of the field and Pajek to identify and depict the key route. The second step is to perform the Global Citation Score analysis (GCS). This selects seminal works (especially the oldest), which, even if not included in the citation network, have a considerable number of citations in the whole Scopus database. The GCS is then integrated with the Global-Local Citation Score (GLCS), a similar analysis performed through the Scopus database that gives relevance also to the most recent papers which had fewer chances to be cited.

Representing a research field only on the basis of the citations of papers could produce biased results since some studies may be not cited for a series of reasons, even if their contribution is relevant. For this reason, it is better to combine the results obtained from the analysis of the citation network with other bibliometric tools, such as the Keywords Network Analysis (KNA) that identifies the co-occurrences around the same keyword or a pair of words that may correspond to a research theme. The KNA is helpful in determining the presence of patterns and trends in a research field (Ding et al., 2001). Moreover, the authors' keywords can be considered a proxy of the papers' content (Strozzi et al., 2017). This is the third and final step of the BNA. In this context, the authors used firstly VOSviewer to compute the co-occurrence of authors' keywords and obtain a network of words instead of citations. Then, a more advanced analysis was performed with the support of Sci2 Tool to run Kleinberg's algorithm and detect the presence of eventual bursts in the authors' keywords over time.

First Phase of SLNA Methodology: Systematic Literature Review

Scope of the Analysis
The scope of this chapter is to conduct a literature review on the topic of Intellectual Capital Disclosure (ICD).

In general, a literature review could have different purposes: some examples are the examination of old theories, the representation of the state of the art on some topics and the provision of suggestions for future investigations (Petticrew & Roberts, 2008). A variety of methods could be applied that differ in terms of the rules to follow for developing the literature review. For instance, it is possible to distinguish between traditional literature reviews (without specific rules) and systematic literature reviews or structured literature reviews, both based on rigid protocol and rules (Massaro et al., 2016).

A recent problem common to several fields of research is the presence of a huge number of publications that make the study of a phenomenon or of a topic harder because of the absence of clarity in previous contributions. Consequently, there is the need for new methodologies helpful in screening the existing literature and selecting the most relevant contributions to scientific research (Colicchia & Strozzi, 2012).

To solve this problem, the SLNA method was applied to the topic of ICD, which is not a new topic, but one of the most debated. It has received considerable attention in the literature, but findings are fragmented and sometimes contradictory. Thus, a need to study the ICD field in a systematic way emerged (Edvinsson, 2013, p. 163), as well as the need to clarify the state of the art of previous contributions with the aim to suggest new possible research trends.

Locating Study
The step that follows the definition of the scope of the analysis is the identification of a set of search strings. In this case, the authors focused on the most common terms that represent the area, i.e. "IC", the three main categories of IC and "disclosure". Then, the keywords included in some relevant papers published by major journals ranked as ABS 3 or 4 (e.g. Abeysekera & Guthrie, 2004; Vergauwen et al., 2007; Li et al., 2008; Chen & Sharma, 2013; Enache & Srivastava, 2018) confirmed the suitability of those terms to represent the field with the correct grade of depth. This part of the analysis is crucial since the results may completely change if different search strings are selected. Moreover, there is the need to reduce individual bias.

The analysis was performed using Scopus: the words ("intellectual capital" OR "relational capital" OR "human capital" OR "structural capital" OR "organizational capital") AND "disclosure" were looked for in "article title, abstract and keywords". No other keywords were added in the analysis to avoid the risk of causing a reduction of the number of papers by restricting the locating study more (Colicchia & Strozzi, 2012; Strozzi et al., 2017). This way allowed us to obtain a better result in terms of the number of topics and trends in the research field.

Study Selection and Evaluation
The search was performed in April 2022. The following criteria were considered to include/exclude papers from Scopus: the analysis was focused on articles and reviews published in English (without any limit as to starting date in order to consider the entire evolution of the topic since the beginning) in the subject areas of "Business, Management and Accounting", "Social Sciences" and "Economics, Econometrics and Finance". These are the research fields in which it is possible to find studies on IC and/or ICD in Scopus. No limits were set in terms of publication years, in order to analyse the whole field and define its temporal

horizon automatically through the set of papers that resulted after the assessment. As a result of the above-mentioned selection process, 426 papers were selected. After this selection process, through the reading of titles and abstracts, some papers not strictly related to our research topic were excluded. The main parameter of exclusion was ICD being a marginal topic in the paper instead of the key issue of analysis; in details, the greatest part of the studies removed from the sample regarded IC in a broader sense and did not focus on its disclosure. The final sample thus results in 361 papers, determining the starting point of the next phase.

Second Phase of SLNA Methodology: Bibliographic Network Analysis

With the first phase (SLR) the most relevant papers were identified. Then, these papers were included in the CNA in order to analyse the development process of the ICD field.

Citation Network Analysis
A "citation network" can be defined as a network with a series of nodes. The nodes are articles connected by links, which represent the citations between the papers and are characterized by arrows. The direction of the arrows goes from cited to citing articles and, thus, by observing it we obtain the evidence of the chronological flow of knowledge from the older papers to the more recent ones (Strozzi et al., 2017). In general, a network is made up of some connected components and some isolated nodes; watching the largest component of the research, it is possible to define the core of the network and to detail the isolated nodes (Fig. 1.1).

With a citation network it is possible to study the data from two different points of view:

1. Analysis of the citation network (static perspective): considering the year of publication it is possible to note if the number of articles increases or not during the period (1999–2022) and thus if the area is in expansion. In this case, the number of papers increases over the period with a gentle fluctuation (Fig. 1.2). It is also possible to identify the journals that published more papers in this field: they are the *Journal of Intellectual Capital* (127 papers), the *International Journal of Learning and Intellectual Capital* (31 papers) and *Corporate Ownership and Control, British Accounting Review,*

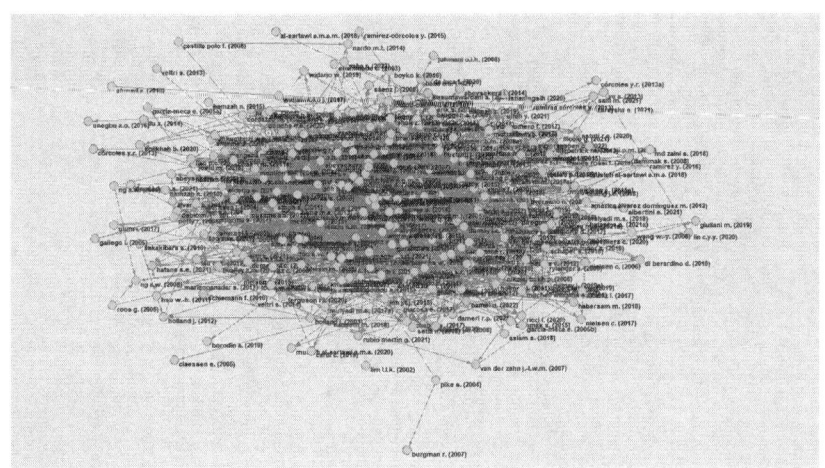

Fig. 1.1 The largest component of the ICD literature network. Source: Processing by Pajek

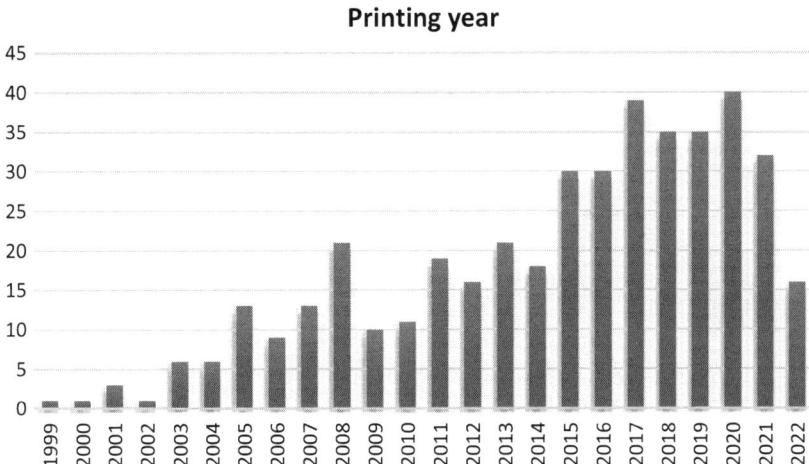

Fig. 1.2 Distribution of scientific articles published during the years 1999–2022. Source: Personal reworking

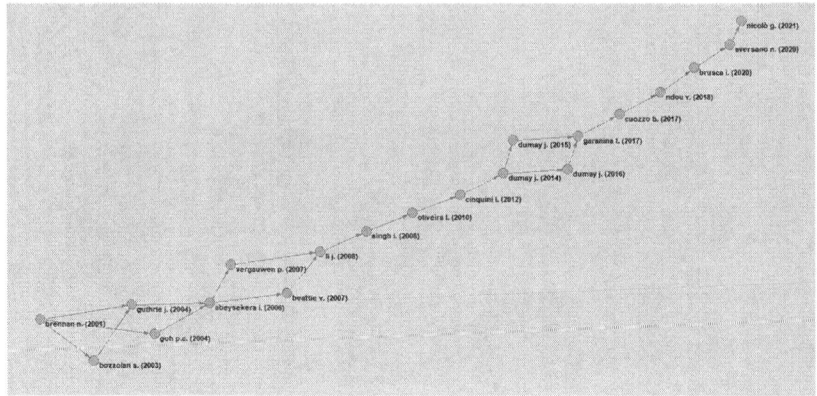

Fig. 1.3 The Main Path of the biggest connected component of the citation network. Source: Processing by Pajek

Academy of Accounting and Financial Studies Journal, *Meditari Accountancy Research* (all with 8 papers). As a result, the predominance of one journal (the *Journal of Intellectual Capital*) in this research field is clear.

2. Main Path Analysis (dynamic perspective).

The first passage of the citation network analysis (CNA) is the exclusion of the isolated nodes in the network because they denote the presence of papers that are neither cited nor citing others. In this analysis, some of the 361 items in the network are not connected to others: the biggest set of connected items consists of 321 papers. Thus, the following steps of the SLNA focused only on this connected component (see Fig. 1.3).

Main Path Analysis
With the aim to investigate the presence of trends in the evolution of the papers' contents, the Main Path analysis was performed (Lucio-Arias & Leydesdorff, 2008). This helps to observe the dynamic perspective of a set of connected papers, and it identifies the most relevant papers published at different times. It is a sort of backbone of the research field (Lucio-Arias & Leydesdorff, 2008; De Nooy et al., 2011).

The process of obtaining the Main Path follows two steps: (1) quantifying the citation traversal weights (or the extent to which a particular citation is necessary to link articles) with the Search Path Count. The weight of the citation is given by the ratio between the number of paths including the citation and the total number of paths between the sources (i.e. articles that do not cite any others) and sinks (i.e. articles that are not cited by any others); (2) extracting the Main Path that identifies the main streams of the considered literature, which is, in our case, a set of 20 papers.

In Fig. 1.3 the Main Path of the biggest component is presented, with evidence of the citation links among the various contributes; Table 1.1, instead, shows further details about each article or review included in the Main Path. The papers range from 2001 to 2021 and their main subject is Intellectual Capital Disclosure (ICD). Two arrows come from other nodes in the case of four papers (Guthrie et al., 2004; Abeysekera, 2006; Li et al., 2008; Garanina & Dumay, 2017). This is usually typical of a literature review or theoretical/methodological suggestions on previous contributions, but in two cases (Li et al., 2008; Garanina & Dumay, 2017) this is not true. A possible explanation could be a citation phenomenon or the fact that those papers are considered milestones in the development of this field of research.

Overall, the findings of our analysis show the presence of two directions in the development of the research field. The first one has an empirical nature and wants to investigate the quality level of the disclosure and the determinants of both the quality and diffusion of ICD, i.e. the extent of ICD use among companies. The second one refers to the critique on methodology used by prior literature in the field and suggestions from future improvements.

In the first literature stream (the study of the determinants, the quality and the diffusion of ICD) there are 13 papers. The oldest is written by Brennan (2001) that studied the extent of Intellectual Capital reporting of a sample of knowledge-based Irish-listed companies; the author examined through a content analysis their Annual Reports and found a low level of disclosure, even though the level of IC assets was high in nearly all the cases.

Then, some authors have deepened the study of the ICD considering separately the three kinds of Capital that compose it: the External (customer-related) Capital, the Internal (structural) Capital and the Human Capital. For instance, Bozzolan et al. (2003) examined the ICD provided by a sample of listed Italian companies in Annual Reports, separately for the three main categories of IC. They observed an extensive disclosure of External Capital and an absence of effect of industry and size

Table 1.1 Main Path papers, with evidence of the title, author, journal, number of citations and year of each publication

Title	Author	Journal	Year	GCS	Type (Empirical or Theoretical)
"Reporting intellectual capital in annual reports: Evidence from Ireland"	Brennan, N.	Accounting, Auditing & Accountability Journal	2001	247	E
"Italian annual intellectual capital disclosure: An empirical analysis"	Bozzolan, S. Favotto, F. Ricceri, F.	Journal of Intellectual Capital	2003	328	E
"Using content analysis as a research method to inquire into intellectual capital reporting"	Guthrie, J. Petty, R. Yongvanich, K. Ricceri, F.	Journal of Intellectual Capital	2004	596	T
"Disclosing intellectual capital in company annual reports: Evidence from Malaysia"	Goh, P.C., Lim, K.P.	Journal of Intellectual Capital	2004	140	E
"The project of intellectual capital disclosure: Researching the research"	Abeysekera, I.	Journal of Intellectual Capital	2006	106	T
"Intellectual capital disclosure and intangible value drivers: An empirical study"	Vergauwen, P., Bollen, L., Oirbans, E.	Management Decision	2007	79	E
"Lifting the lid on the use of content analysis to investigate intellectual capital disclosures"	Beattie, V., Thomson, S.J.	Accounting Forum	2007	257	T

(*continued*)

Table 1.1 (continued)

Title	Author	Journal	Year	GCS	Type (Empirical or Theoretical)
"Intellectual capital disclosure and corporate governance structure in UK firms"	Li, J., Pike, R., Haniffa, R.	Accounting and Business Research	2008	292	E
"Determinants of intellectual capital disclosure in prospectuses of initial public offerings"	Singh, I., Van Der Zahn, J.-L.W.M.	Accounting and Business Research	2008	64	E
"Intellectual capital reporting in sustainability reports"	Oliveira, L., Rodrigues, L.L., Craig, R.	Journal of Intellectual Capital	2010	52	E
"Analyzing intellectual capital information in sustainability reports: Some empirical evidence"	Cinquini, L., Passetti, E., Tenucci, A., Frey, M.	Journal of Intellectual Capital	2012	49	E
"A review and critique of content analysis as a methodology for inquiring into IC disclosure"	Dumay, J., Cai, L.	Journal of Intellectual Capital	2014	109	T
"Using content analysis as a research methodology for investigating intellectual capital disclosure: A critique"	Dumay, J., Cai, L.	Journal of Intellectual Capital	2015	69	T
"A critical reflection on the future of intellectual capital: from reporting to disclosure"	Dumay, J.	Journal of Intellectual Capital	2016	225	T

(*continued*)

Table 1.1 (continued)

Title	Author	Journal	Year	GCS	Type (Empirical or Theoretical)
"Forward-looking intellectual capital disclosure in IPOs: Implications for intellectual capital and integrated reporting"	Garanina, T., Dumay, J.	Journal of Intellectual Capital	2017	33	E
"Intellectual capital disclosure: a structured literature review"	Cuozzo, B., Dumay, J., Palmaccio, M., Lombardi, R.	Journal of Intellectual Capital	2017	89	T
"Understanding intellectual capital disclosure in online media Big Data: An exploratory case study in a university"	Ndou, V., Secundo, G., Dumay, J., Gjevori, E.	Meditari Accountancy Research	2018	20	E
"Intellectual capital disclosure and academic rankings in European universities: Do they go hand in hand?"	Brusca, I., Cohen, S., Manes-Rossi, F., Nicolò, G.	Meditari Accountancy Research	2020	16	E
"The Integrated Plan in Italian public universities: new patterns in intellectual capital disclosure"	Aversano, N., Nicolò, G., Sannino, G., Tartaglia Polcini, P.	Meditari Accountancy Research	2020	8	E
"ICD corporate communication and its determinants: evidence from Italian listed companies' websites"	Nicolò, G., Aversano, N., Sannino, G., Tartaglia Polcini, P.	Meditari Accountancy Research	2021	6	E

on the content disclosed, while the same two variables influenced the amount of information reported. A similar contribution has been developed by Goh and Lim (2004), which examined through a content analysis the Annual Reports of the top 20 profit-making public-listed companies in

Malaysia, distinguishing among Internal Capital, External Capital and employee competences. The authors, in this case, noticed a high incidence of ICD in the reports from a qualitative point of view with high relevance of External Capital as Bozzolan et al. (2003), even though there was lack of voluntary quantitative information in financial statements. The same research settings—although with different outcomes—were kept by Vergauwen et al. (2007), which instead analysed the extent of ICD in the Annual Reports of Swedish, British and Danish companies, in order to measure the three categories of Human, Structural and Relational Capital. The results in this case were focused on Structural Capital, suggesting that firms with higher level of such possessions disclosed more intangible information in their reports; regarding Human and Relational Capital, instead, there were no significant connections with ICD. Cinquini et al. (2012) followed the same research path by analysing 37 sustainability reports published by Italian-listed companies, in order to evaluate the content, frequency and quality of ICD. After a content analysis, the authors defined an increase of ICD incidence over time, regarding mainly the Human Capital; moreover, ICD proved to be mostly quantitative but non-financial and not time-oriented.

Other studies focused on the determinants of the ICD quality and their adoption. Li et al. (2008) showed that some corporate governance variables (such as board composition, share concentration, audit committee size and frequency of audit committee meetings) positively affect ICD in a sample of 100 UK-listed firms. The same year, Singh and Van der Zahn (2008) investigated the relation among ICD and three determinants—such as share retention, proprietary costs and governance structure—after examining the reports of 444 initial public offerings (IPOs) listing companies in the Singapore Stock Exchange (1997–2006). The results highlighted a significant positive link between ICD and share retention, and a negative effect of proprietary costs on the previous relation. Then, Oliveira et al. (2010) measured the determinants of ICD in sustainability reports of Portuguese firms for year 2006 through an index of intangibles voluntary disclosure; they found out that applying Global Reporting Initiative (GRI) framework and being a listed company are two variables that highly influence ICD. Garanina and Dumay (2017) investigated the extent to which managers and owners of technology companies listed on the NASDAQ disclosed IC in initial public offering (IPO) prospectuses for the period 2002–2013. Overall, the IPO prospectuses contained significant amounts of ICD, with an increase after the global financial crisis and a greater positive effect on post-issue stock performance.

More recently, some authors focused specifically on universities' disclosure. The paper of Aversano et al. (2020) investigated the Integrated Plans developed by 60 Italian public universities for years 2018–2020 to define the level, form and location of ICD; they confirmed the higher level of disclosure of Human Capital—while the overall level of the three categories' disclosure was medium—the prevalence of quantitative information and the focus on the parts related to the strategic framework and organizational performance. Moreover, attention paid to the new online source of disclosure has increased: for instance, Ndou et al. (2018) analysed how information published on online media could be a relevant source of data for ICD in universities. Social media information confirmed to be an essential means to communicate IC and understand how it is integrated in the organization. This study is similar to the research of Brusca et al. (2020). They compared the ICD made by 128 different universities in three European countries, finding a non-homogeneous disclosure level, with higher results for big universities according to World Ranking. They observed also that almost the whole communication referred to Human and Internal Capital and identified the presence of a positive correlation between IC web disclosure and the academic ranking. Finally, Nicolò et al. (2021) analysed the three categories of capital in relation to the academic performance of 59 Italian public universities through the content of their performance reports. The outcomes regarded mainly Human Capital, whose disclosure had a high value for the sample; however, in general, when performance was higher, the universities gave more space to IC and its sub-components disclosure in their reports.

In summary, this first research stream is focused on the ICD extent, its quality and the identification of its main determinants. Up to now, it has been possible to observe that the amount of the information disclosed on IC is still low, as is also its quality, even if there is an increasing trend at least regarding the quantity of the information disclosed. The IC category that is considered more frequently by companies is External Capital, while less attention is paid to Structural and Human Capital. Instead, if we consider the disclosure provided by universities, almost all the attention is given to Human Capital. Finally, more recently researchers have shifted their focus from the traditional types of disclosure (e.g. Annual Reports) to the new online source of disclosure which may improve the ICD level and quality.

Seven papers belong to the second stream of literature, i.e. the critique on methodology used by prior literature in the field and suggestions from future improvements. Four studies (Guthrie et al., 2004; Beattie & Thomson, 2007; Dumay & Cai, 2014, 2015) aimed at reviewing and criticizing the use of content analysis for the assessment of ICD. Firstly, Guthrie et al. (2004) offered useful observations and issues on the practical use of the content analysis technique, since the data collected are often uncomparable because of their context or location of origin. Then, Beattie and Thomson (2007) highlighted the absence of a univocal definition of the IC concept and the lack of a common language to define the extent and nature of the disclosure provided. They pointed out that this technique needs more transparency, reliability and a lower level of subjectivity, since there underlies too much space for interpretation and lacks comparability. Dumay and Cai contributed to the research field with two different publications. In the first one (2014) the authors analysed 110 articles that used content analysis to study ICD and concluded that Annual Reports—even though highly used—are not the best source to research and find new evidence about ICD; their thought about this topic is not very positive, since Annual Reports have only confirmed that companies are not willing to disclose IC information to their stakeholders. In their second contribute (2015)—consistent with the previous studies of Guthrie et al. (2004), Abeysekera (2006) and Beattie and Thomson (2007)—the authors went further and demonstrated the inconsistency of application of content analysis in ICD field. Specifically, the analysis was carried out without generally accepted approaches to define research questions, the setting precluded the comparability of findings and replicability of the studies, and there was lack of reliability and validity.

Other authors have also contributed to this stream of research in a more general form. For instance, Abeysekera (2006) illustrated some of the strengths, weaknesses and gaps in the extant research on IC, such as the lack of coherence for the presence of many different definitions of IC and ICD (which do not sufficiently address the issue of value creation) and the absence of a shared framework for analysing Annual Reports, which causes a problem of comparability between previous ICD studies (similarly to Guthrie et al., 2004; Beattie & Thomson, 2007; Dumay & Cai, 2014, 2015). Then, Dumay (2016) continued his research line by suggesting researchers to focus on how companies disclose previously hidden information instead of their reporting. Finally, Cuozzo et al. (2017) provided a literature review of ICD and identified some future research themes and the evolution of the topic. They highlighted a lack of innovation in the evolution of the ICD research field and, consistent with previous publications findings of Dumay and Guthrie, confirmed that academic studies were still lying on the surface of reports, talking about general issues, instead of investigating practically ICD at organizational level. For this reason, the authors suggested that researchers needed to conduct future research into how organizations use intangibles to create value.

To summarize, the second stream of literature provides a critique of previous IC research concerning the methodologies applied (e.g. the problem of subjectivity of the content analysis technique, absence of a common definition of IC and ICD and of a shared framework for conducting the analysis), with a strong focus on content analysis issues for academic research, as it was previously carried out. Moreover, those studies give suggestions to the future of the research field which needs to focus more on how the information is disclosed by companies and less on reporting practices. However, all the authors agreed that the main motivations of companies for disclosing IC information are the needs to gain legitimacy and to reduce the information asymmetry with their stakeholders.

Global Citation Score Analysis
Considering only the Main Path could result in a loss of some contributions in the field of the ICD. As explained before, some papers may be excluded from the citation network for the absence of citations linking them to other studies.

To compensate for this limitation, additional analyses were performed, like the Global Citation Score (GSC). The GCS is based on the global number of citations in the whole Scopus database. A higher number of citations could be a proxy for the influence of a paper into a specific stream of literature (Knoke & Yang, 2008), even if it is not necessarily a synonym of high-quality research (Dawson et al., 2014). We also have to consider that, typically, the oldest articles are characterized by a higher number of citations with respect to recent studies. For all these reasons, it is important to combine the two above-mentioned techniques (CNA and GCS), because only if taken together they can solve the problem posed by their own limits.

Table 1.2 reports the ten most cited papers ranked according to their GCS, equal to the total number of citations in Scopus. Here, six papers are already included in the Main Path and this is a confirmation of the

Table 1.2 Global Citation Score (GCS) of the ten most cited papers, together with title, authors, journals and years of publication

Rank	Title	Author	Journal	Year	GCS	Main Path
1	"Using content analysis as a research method to inquire into intellectual capital reporting"	Guthrie, J. Petty, R. Yongvanich, K. Ricceri, F.	Journal of Intellectual Capital	2004	596	x
2	"Italian annual intellectual capital disclosure: An empirical analysis"	Bozzolan, S. Favotto, F. Ricceri, F.	Journal of Intellectual Capital	2003	328	x
3	"Intellectual capital disclosure and corporate governance structure in UK firms"	Li, J. Pike, R. Haniffa, R.	Accounting and Business Research	2008	292	x
4	"The voluntary reporting of intellectual capital: Comparing evidence from Hong Kong and Australia"	Guthrie, J. Petty, R. Ricceri, F.	Journal of IntellectualCapital	2006	266	
5	"Lifting the lid on the use of content analysis to investigate intellectual capital disclosures"	Beattie, V. Thomson, S.J.	Accounting Forum	2007	257	x
6	"Reporting intellectual capital in annual reports: Evidence from Ireland"	Brennan, N.	Accounting, Auditing & Accountability Journal	2001	247	x
7	"Exploring the effects of corporate governance on intellectual capital disclosure: An analysis of European biotechnology companies"	Cerbioni, F. Parbonetti, A.	European Accounting Review	2007	240	

(*continued*)

Table 1.2 (continued)

Rank	Title	Author	Journal	Year	GCS	Main Path
8	"A critical reflection on the future of intellectual capital: From reporting to disclosure"	Dumay, J.	*Journal of Intellectual Capital*	2016	225	x
9	"Is intellectual capital performance and disclosure practices related?"	Mitchell Williams, S.	*Journal of Intellectual Capital*	2001	215	
10	"Disclosure of information on intellectual capital in Danish IPO prospectuses"	Nikolaj Bukh, P. Nielsen, C. Gormsen, P. Mouritsen, J.	*Accounting, Auditing & Accountability Journal*	2005	206	

reliability of this kind of analysis in identifying the most relevant studies in a research field. However, there are additional papers not included in the Main Path analysis.

Overall, these papers confirm the presence of the two streams of research already identified through the Main Path analysis: in addition to the papers already analysed for the Main Path, all the new contributions studied the determinants of the quality and the diffusion of ICD (Williams, 2001; Bukh et al., 2005; Guthrie et al., 2006; Cerbioni & Parbonetti, 2007). Starting from the most cited of those studies, Guthrie et al. (2006) analysed voluntary ICD of listed companies based in Australia and Hong Kong; the authors, as previous studies, confirmed that ICD was better performed by bigger companies, even though the overall level of voluntary disclosure was low and consisted mainly in qualitative information. Then, Cerbioni and Parbonetti (2007) examined a sample of European biotechnology firms to confirm that there exists a relationship between corporate governance variables and ICD quality. Specifically, they defined that the proportion of independent directors affects positively the disclosure of information on the internal corporate governance structure, e.g. the board composition significantly improves the readability of Annual Reports; on the contrary, the CEO duality prevents the forward-looking information disclosure. Williams (2001), instead, analysed the Annual

Reports of 31 FTSE 100 listed companies for the period 1996–2000 and focused on ICD quantity, defining that there was lack of systematic relationship between IC performance and its disclosure extent; moreover, it seemed that companies perceived ICD as a form of losing competitiveness when IC performance was very high. Instead, leverage, industry exposure and listing status were significant variables influencing ICD quantity. Finally, Bukh et al. (2005) examined the voluntary disclosure included in Danish IPO prospectuses for the period 1999–2001 to define whether the IC was communicated and the determinants of its extent. The authors found out that the managerial ownership extent before the IPO and the sector significantly influenced the quantity of ICD, while other variables regarding the company (such as its age and size) had neutral effects on voluntary disclosure.

Moreover, the results are integrated with a Global-Local Citation Score (GLCS) analysis, based on the total number of citations in the Scopus database but only related to the most recent papers (published during the period 2019–2022). The aim of the GLCS is overcoming the main bias of the GCS, i.e. the recognition of older seminal papers by the academics in the field as relevant for their high number of citations. This leads to a higher number of citations and greater emphasis on the GCS compared to other recent studies, regardless of their quality and importance for future developments. Hence, the GLCS gives relevance to the papers that had fewer chances of being cited, focusing only on the last years of development of the research field.

Table 1.3 reports the ten most cited papers published from 2019 to 2022. None of them is included in the Main Path.

Also these studies confirm the two streams of research discussed before in the Main Path analysis (four over ten studies are theoretical, while the others are empirical), but a higher attention to the recent topic of Integrated Reporting (IR) and its relationship with the disclosure of Intellectual Capital emerges in this context (Dumay et al., 2019; Beretta et al., 2019; de Villiers & Sharma, 2020; Salvi et al., 2020a, 2020b; Vitolla et al., 2020). For instance, de Villiers and Sharma developed the most cited contribute (2019), in which reported some reflections on different frameworks and standards available to companies willing to disclose IC information using international schemes, such as IR and GRI frameworks. It has emerged that there is not a particular form that includes at best all IC information; therefore, their integration with the use of financial statements and potential future developments will explain well IC contents and

Table 1.3 Global-Local Citation Score (GLCS) of the ten most cited papers published in 2022, 2021, 2020 and 2019, together with title, authors and journals

Rank	Title	Author	Journal	Year	GCS
1	"A critical reflection on the future of financial, intellectual capital, sustainability and integrated reporting"	de Villiers, C. Sharma, U.	Critical Perspectives on Accounting	2020	62
2	"Developing trust through stewardship: Implications for intellectual capital, integrated reporting, and the EU Directive 2014/95/EU"	Dumay, J. La Torre, M. Farneti, F.	Journal of Intellectual Capital	2019	56
3	"Critical mass of female directors, human capital, and stakeholder engagement by corporate social reporting"	Amorelli, M.-F., García-Sánchez, I.-M.	Corporate Social Responsibility and Environmental Management	2020	47
4	"Does board gender diversity influence voluntary disclosure of intellectual capital in initial public offering prospectuses? Evidence from China"	Nadeem, M.	Corporate Governance: An International Review	2020	37
5	"Intellectual capital disclosure in integrated reports: The effect on firm value"	Salvi, A., Vitolla, F., Giakoumelou, A., Raimo, N., Rubino, M.	Technological Forecasting and Social Change	2020a	34
6	"Does environmental, social and governance performance influence intellectual capital disclosure tone in integrated reporting?"	Beretta, V. Demartini, C. Trucco, S.	Journal of Intellectual Capital	2019	31
7	"Intellectual capital and the firm: Evolution and research trends"	Martín-de Castro, G. Díez-Vial, I. Delgado-Verde, M.	Journal of Intellectual Capital	2019	26

(*continued*)

Table 1.3 (continued)

Rank	Title	Author	Journal	Year	GCS
8	"The role of board of directors in intellectual capital disclosure after the advent of integrated reporting"	Vitolla, F., Raimo, N., Marrone, A., Rubino, M.	*Corporate Social Responsibility and Environmental Management*	2020	24
9	"Social media disclosure of intellectual capital and firm value"	Musleh Al-Sartawi, A.M.A.	*International Journal of Learning and Intellectual Capital*	2020	22
10	"Does intellectual capital disclosure affect the cost of equity capital? An empirical analysis in the integrated reporting context"	Salvi, A., Vitolla, F., Raimo, N., Rubino, M., Petruzzella, F.	*Journal of Intellectual Capital*	2020b	21

will suit fairly the ICD needs of companies. Dumay et al. (2019) explored how IC and IR can be combined to develop a new extended model of disclosure for companies to comply with EU Directive 2014/95/EU. Specifically, the authors spotted lack of consistency between reporting and corporate behaviour and defined that disclosing more information was not a solution when no practical improving was made in resources management; instead, managers should apply the new model which abandons the "bonus contract" of the agency theory application. Then, Beretta et al. (2019) analysed the connection between ICD in IR and companies' performance after having examined all the reports developed by European-listed firms in the period 2011–2016. They obtained some evidence regarding the features of ICD: it was mostly colloquial, written with a "positive tone", backward-oriented and little evidence on Human Capital. Moreover, the more optimistic was the tone, the higher were ESG performance. Salvi et al. wrote two important contributions in this specific research stream (2020a, 2020b). In the first one, the authors developed an empirical analysis of 110 companies which developed IR to understand how the ICD quality influenced the firm value; the outcomes showed that all the three categories of IC conditioned positively the firm's value, creating positive externalities for both internal and external stakeholders (included investors and regulators). In the second one, instead, Salvi and the other authors focused on the influence of ICD on the cost of equity capital. After examining 164 IR, they defined that a higher level of disclosure has a positive impact on cost of equity capital, causing its decrease. Finally, Vitolla et al. (2020) used a sample of 130 heterogeneous international firms which developed IR to focus the attention on the Board of Directors' (BoD) influence on ICD quality. This study identified a new system to measure the quality of ICD and used it to define the features of BoD which influence positively the IR quality, such as the size, activity diversity and independence. Apart from IR research line, the ICD theme is included in other four contributions, including two which are focused specifically on BoD gender diversity (Amorelli & García-Sánchez, 2020; Nadeem, 2019). The first study, carried out by Amorelli and García-Sánchez (2020), aimed to explain whether subsists a relation between the presence of females in the BoD and Corporate Social Responsibility disclosure, as well as the influence of the BoD components features meant like parameters to measure Human Capital. Analysing more than 9700 companies between 2007 and 2016 allowed the authors to define that 3 women in the BoD are the minimum critical mass to generate a positive effect on CSR disclosure but also that great BoD skills, background and experience have a good impact too. Similarly, the research of Nadeem (2019) highlighted a positive association between BoD diversity and ICD, with greater effects when at least two women are in the Board, although females generate a negative impact on disclosure if are independent directors. Finally, two papers do not concern the topics just identified. Martín-de Castro et al. (2019) conducted a literature review of previous contributions on the IC literature, finding four main research streams for future publications such as ICD, measurement, IC's role in HR and social capital practices, and IC in new business models. Musleh Al-Sartawi (2021), instead, examined Kuwaiti and Omani listed firms to define that the social media disclosure of IC influences significantly and positively the value of firms.

Co-occurrence Analysis of Authors' Keywords
The above-mentioned analyses (CNA and GCS) focus mainly on the papers included in the connected component and on the most cited papers. The results of the literature review can be improved with a study of the author keywords network (Ding et al., 2001) on the set of papers resulting from the SLR process, also including the isolated nodes of the connected component. At the basis of the co-occurrence analysis there is

the assumption that authors' keywords are a proxy for the papers' content (Strozzi et al., 2017). Compared to citation analysis, in which the abstract and full paper are the main source of contents, studying keywords allows to examine the field from a more generic perspective, unrelated to individual papers. Therefore, this analysis is complementary to the previous phases to identify the whole research themes and new trends in the research field (Ding et al., 2001).

The software used is VOSviewer (Van Eck & Waltman, 2010). The first step is the extraction of the authors' keywords from the 361 papers selected in Scopus during the SLR phase. Then, a co-word network was built and analysed through the VOSviewer software, which gives the locations of items on a map. The parameter of the minimum number of occurrence of keywords is selected equal to 4 because a higher value could provoke the exclusion of specific keywords giving relevance only to general terms, while a smaller value would decrease the significance of the analysis by including keywords that are not enough relevant (Strozzi et al., 2017). This step is essential to define a good mid-way criteria of keywords selection and perform as a consequence a proper clustering process that enhances both the main topics and the peculiarities studied frequently by prior literature. Moreover, we used the function of the Thesaurus file from VOSViewer to unify terms that have the same meaning but different spelling. This provides a much more consistent network.

Figure 1.4 shows the results obtained from the analysis of the authors' keywords of the 361 papers extracted from Scopus. A total of 23 keywords have been detected by the VOS algorithm, divided into 5 main clusters. An increasing dimension of the size of a circle means that a keyword is more common among all the papers.

These clusters confirmed not only the research trends identified from the Main Path analysis but also some peculiarities described in the Main Path analysis itself (e.g. the common usage of Annual Reports to perform content analysis, being a listed company as usual feature to become part of the sample, the categories of capitals, the different geographical regions involved in the analysis, etc.). Therefore, the keywords confirm that the contributes can be grouped under the two above-mentioned research streams—each of those with its details and further lines—even though more space is given to issues related to the empirical analysis papers compared to the critical points highlighted by theoretical contributions (literature review). Indeed, Cluster 1 (the most generic and bigger in terms of single keywords popularity) focuses on IC, voluntary disclosure, the

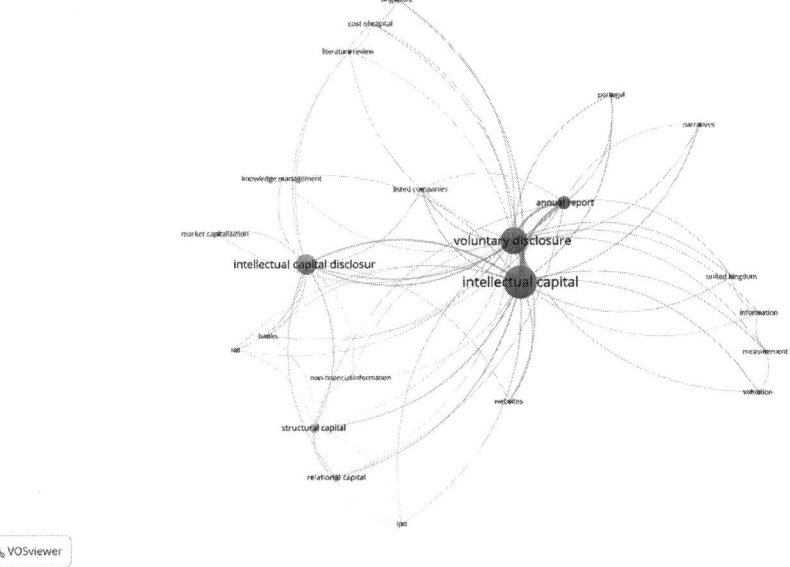

Fig. 1.4 Resulting clusters of the co-occurrence network of author keywords analysis, based on algorithm of the software VOSviewer. The colours red, green, blue, yellow and purple represent, respectively, cluster number 1, 2, 3, 4 and 5. Source: processing by VOSviewer

features of ICD and the kind of sample or input data used for the analysis; Cluster 2 includes Intellectual Capital Disclosure and some of its determinants; Cluster 3 reports various ways to manage IC in order to obtain related information to disclose; Cluster 4 shows the different IC categories (in particular, the Relational and the Structural Capital); Cluster 5 represents the paper based on literature reviews and some variables (e.g. the cost of capital). There is evidence about different geographical areas, even if the countries have been allocated to various clusters. Except for some clusters—e.g. the one related to the three categories of capitals or to the determinants of ICD—a couple of them don't represent clearly a specific research stream, Cluster 5 above all.

The author keywords for each cluster are presented in Table 1.4; the keywords present most frequently are shown in bold.

Table 1.4 Author keywords in each of the three identified clusters

Cluster	Keywords
1	Annual Report, **Intellectual Capital, Listed companies**, Narratives, Portugal, **Voluntary disclosure**, Websites
2	Banks, Icd, **Intellectual Capital Disclosure**, Knowledge Management, Market capitalization
3	Information, Measurement, United Kingdom, Valuation
4	Ipo, Non-financial information, Relational capital, Structural capital
5	Cost of Capital, Literature Review, Singapore

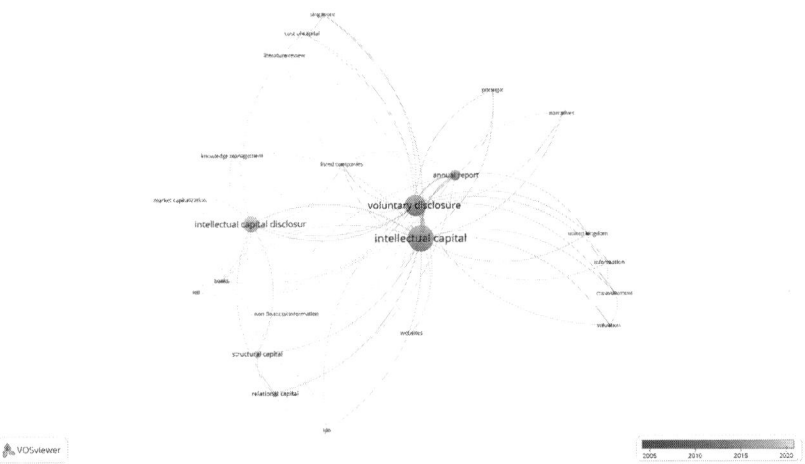

Fig. 1.5 Co-occurrence network of author keywords according to the year of publication. Source: Processing by VOSviewer

In VOSviewer it is also possible to classify the keywords on the basis of the average year of publication (Fig. 1.5): blue indicates the oldest papers, while yellow the most recent ones.

The most recent keywords are: "*Websites*", "*Market capitalization*", "*Non-financial information*". From this analysis, it is possible to obtain information about the trending topics and the most recent. Firstly, it highlights that recent studies used websites information in order to investigate IC information, in addition or substituting the Annual Reports content; therefore websites are not determinants or variables related to ICD, but the input data source to overtake previous content analysis issues. Then, market capitalization represents one of the variables that have been investigated more recently, concerning the effects performed directly by ICD. Moreover, other frequently used keywords in the recent period are *"Structural Capital"* and *"Relational Capital"*, since many researchers analysed the IC information related to the three capitals classification and the most represented category in voluntary disclosure depending on the context examined. They highlight the importance of conducting studies only on a specific category of IC because for each category it is possible to observe different findings, for example how different variables could impact them. Finally, other keywords such as "ICD", "literature review", "knowledge management" and "IPO" have been used quite recently, highlighting different features of the studies.

Kleinberg's Burst Detection Algorithm

Within the literature there is a flow of topics that become popular and debated more or less intensively in a specific period of time. It is possible to name this phenomenon a "burst of activity" (Strozzi et al., 2017), since the use of a keyword occurs at a certain moment, grows more or less over a timespan and then fades away leaving space to new emerging issues.

The first author that used a formal approach for modelling "bursts" is Kleinberg (2003). In this model, bursts represent state transitions.

In this study, Kleinberg's algorithm was applied to the authors' keywords as it was done for the co-occurrence analysis in Sect. 1.2.3.4; the results of the analysis range between year 2002 and 2022. The authors' keywords of all the papers are extracted and pre-processed (normalized) using the Sci2 software.

The results of the application of the burst detection algorithm are shown in Fig. 1.6.

While the width of the bars enhances the intensity of keywords' use in a specific period of time, the length of the bars defines the duration of this phenomena.

It is possible to assert that while in the first ten years of analysis there was a fewer quantity of ongoing trends that lasted many years, in the second half of the period a lot of keywords were used more intensely but for a shorted timespan. For instance, "Annual Report" was very used for a longer period because a great part of the empirical analysis at that time was based on the examination of Annual Reports' contents; on the contrary, topics like "Integrated Reported", "performance" and "governance"

1 THE EVOLUTION AND NEW TRENDS OF THE INTELLECTUAL CAPITAL... 29

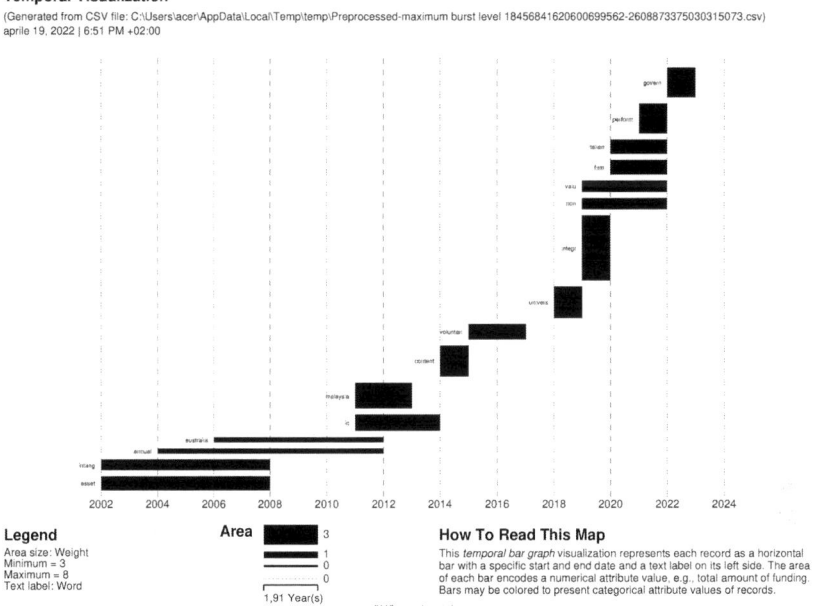

Fig. 1.6 Burst detection algorithm applied to normalized author keywords from 2001 to 2022. Source: Processing by Sci2 software

issues have been investigated more frequently but for a shorter period. According to this analysis, the main concepts dealt by literature over the years refer to intangible assets, their determinants and diverse contexts in which ICD can be studied, with a particular focus on differences among countries. Among the most recent and specific topics, in particular, the governance variables emerged as determinant of ICD, as well as the Italian context and the educational sector, i.e. "universities". The latter proved to be the subject of analysis quite intensely, corroborating what emerged firstly by the citation analysis. Finally, other relevant topics studied recently are the "performance" disclosed through ICD and "Integrated Reporting", which is the new frontier within the "voluntary" reporting tools on the IC.

1.3 Identifying Research Directions in the ICD Field

The study of the dynamic evolution of ICD research field through the citations' analysis showed that even though the literature on this topic is really rich, only two sub-sets of study themes could be identified: the first one refers to the ICD quantity and quality and the identification of its main determinants, while the second one aims at the construction of a theoretical framework for further investigation into this topic or provides some technical suggestions related to the presence of methodological issues. This was defined at the Main Path analysis step and was further validated by the GCS and GLCS phases. The following studies on keywords, however, brought out different clusters to describe the overall field and added up many specific facets of literature. On the one hand, this confirmed the utility of using a multi-analysis methodology to depict comprehensively a research field. On the other hand, having a big picture allowed us to identify some key themes and, mostly, further development and directions for researchers, in particular regarding some methodological changes that are recommended throughout literature. Specifically, six main themes can be identified: "Data source", "Reporting vs. Disclosure", "Static vs. dynamic studies", "Comparative studies", "IC as a whole" and "Methodology". For each theme future research directions are suggested.

Regarding the first one ("Data source"), new studies should refer to new data sources (e.g. website, a new form of reporting such as Integrated Reporting) and new theories should be applied (Garanina & Dumay, 2017; Ndou et al., 2018; Dumay et al., 2019; Beretta et al., 2019; Salvi et al., 2020a, 2020b; Aversano et al., 2020; Brusca et al., 2020; De Villiers & Sharma, 2020; Vitolla et al., 2020; Musleh Al-Sartawi, 2021). Otherwise, as suggested by Dumay and Cai (2014, 2015) the increase of knowledge on this topic will remain only incremental. Moreover, researchers must justify their choices in terms of data sources and theories, have to answer pertinent research questions and clarify the boundaries of their analysis in order to make the studies replicable (Beattie & Thomson, 2007; Dumay & Cai, 2015).

The second theme ("Reporting vs. Disclosure") implies considering internal IC practices and documents such as interviews, field notes, tweets, blogs and news media in order to conduct a day-by-day analysis. Researchers, as companies, have to shift their focus from reporting ("detailed periodic account of a company's activities, financial condition, and prospects that is made available to shareholders and investors") (Dumay, 2016: 178) to disclosure ("the revelation of information that was previously secret or unknown") (Dumay, 2016: 178) and consider that the disclosure does not necessarily imply the detection of sensitive information (Dumay, 2016; Garanina & Dumay, 2017). It is possible to define a strong link with the previous theme, since the source of data (e.g. Annual Report vs. Social Media contents) affects significantly the input of research, whether it is related more comprehensively to disclosure or solely to reporting.

The third and fourth themes ("Static vs dynamic studies" and "Comparative studies") concern the need to conduct more longitudinal studies and cross-national comparison (Brennan, 2001; Bozzolan et al., 2003; Bukh et al., 2005 Guthrie et al., 2006; Goh & Lim, 2004; Li et al., 2008; Cerbioni & Parbonetti, 2007; Brusca et al., 2020; Aversano et al., 2020; Nicolò et al., 2021) to monitor the progress and the development of ICD practices and to understand which contextual variables affect it. In fact, geographical features could strongly influence the results of empirical analysis, when the sample is focused solely on a specific country (e.g. because of cultural, political, economic, social, legal, environmental variables). In order to validate accuracy of results and their generalization, it is better to perform tests in other environments, especially when the country considered initially has peculiarities that can be influent.

Regarding the fifth theme ("IC as a whole"), future research should study the three different categories of IC (Organizational, Human and Relational Capital) separately because this could lead to different and more precise findings and insights (Bozzolan et al., 2003; Goh & Lim, 2004; Vergauwen et al., 2007; Cinquini et al., 2012; Aversano et al., 2020). In fact, different contexts and environments (e.g. universities vs. manufacturing or services companies) could enhance different IC categories with significant consequences on the related disclosure too.

Finally, the sixth theme ("Methodology") refers to the research methods: since the use of content analysis implies a series of disadvantages (e.g. subjectivity, lack of transparency and reliability, difficulties of replicability in different contexts or generating deeper knowledge in this field of research), the rigorousness of this methodology application can be affected. Therefore, it could be useful to test other methods, such as questionnaire survey, interviews and case studies (Guthrie et al., 2004; Beattie & Thomson, 2007; Dumay & Cai, 2014, 2015), or to use several methods at the same time to overcome the limits of the more traditional content analysis. Moreover, since content analysis was mainly applied through Annual Reports or voluntary reporting examination, this issue can be significantly linked to themes 1 and 2. Therefore, when performed, content analysis should take into account other sources which are not voluntarily disclosed by the company, such as the website or social networks, and try to figure out the practices which are carried out by the firms but not communicated (Dumay, 2016; Nicolò et al., 2021).

In Table 1.5 the evolution of the key concepts of ICD is summarized, along with future research directions.

Table 1.5 Research directions in the ICD field from a methodological point of view

Main themes	From	To	Research directions
1. Data source	Annual report	New data sources (e.g. website, Integrated Reporting)	Consider more than just one data source for a complete picture of the companies' ICD
2. Reporting versus disclosure	Voluntary disclosure	Involuntary disclosure	Analyse not only the external IC (the information directly published by companies) but also the internal IC practices to conduct a day-by-day analysis
3. Static versus dynamic studies	Studies of just one year	Longitudinal studies	Expand the horizon considered in the analysis to identify possible ICD trends
4. Comparative studies	Studies in the same setting	Cross-national studies	Conduct a comparison of different contexts to understand which setting positively affects the ICD
5. IC as a whole	IC concept	Three IC categories	Analyse the three different IC categories (organizational, human and relational capital) separately for more detailed results on the ICD quality and diffusion
6. Methodology	Content analysis	Content analysis + other techniques	Enlarge the number of technical tools used to conduct the study to have more reliable and complete results

1.4 Discussion and Conclusions

The present literature review is conducted with the aim of providing a comprehensive view of the state of the art of Intellectual Capital Disclosure (ICD) and suggesting some possible new research directions, answering the call of prior literature both in terms of contents (Edvinsson, 2013; Dumay & Cai, 2014; Afeltra et al., 2022) and methodologies (Colicchia & Strozzi, 2012; Dumay, 2014; Massaro et al., 2016; Afeltra et al., 2022).

Specifically, ICD has received considerable attention in the literature, but previous contributions are fragmented and reported contradictory findings. Following these premises, there is the need to systematize this research topic (Edvinsson, 2013, p. 163) and to hypothesize new trends for research (Colicchia & Strozzi, 2012).

A SLNA analysis was performed to obtain a dynamic representation of the evolution of the flow of knowledge on this topic. It combines the systematic literature review (SLR) and the citation network analysis (CNA). In particular, the Main Path analysis helped in localizing the seminal works that represent a reference point for recent studies (De Nooy et al., 2011; Lucio-Arias & Leydesdorff, 2008; Strozzi et al., 2017), and, consequently, it was useful for identifying the evolution of the research field, while the analysis of keywords suggested the most relevant contributions in the field. In addition, other tools were used (such as the Global Citation Score analysis and the keyword analysis through Kleinberg's algorithm) in order to cope with some limitations of the SLNA (e.g. the exclusion of recent papers because of the lower amount of citations).

The purpose of the chapter is to try to define the main themes that have been developed within ICD research and future research.

The findings of the SLNA show that ICD is a quite debated topic, both theoretically and empirically. On the one hand, there is a high amount of empirical literature that aims at defining the ICD extent, quality and its main determinants through the validation of empirical frameworks after the analysis of companies' disclosure contents. On the other hand, more theoretical literature is focused on the conceptualization of new frameworks and on the critique of the methodologies currently used by authors in empirical research.

More specifically, the current work suggests the presence of two different directions in the development of the research field.

The first one, being related to the ICD quality, diffusion and the identification of its drivers, shows that the level of ICD is still low, but there is an increasing trend in the quantity of the information disclosed, being companies and their information systems affected by some variables (e.g. corporate governance characteristics). In this case, industry and country variables can influence the results of the analysis, since some sectors have been highly examined compared to others (e.g. educational industry).

The second stream, having theoretical nature, provides both updated conceptual frameworks to support more accurate empirical investigations on the topic of ICD as well as technical suggestions from a methodological point of view (e.g. critique of the content analysis that implies a high level of subjectivity, absence of a shared definition of IC and ICD). Therefore, it is possible to state that the research streams are highly correlated and that, following the theoretical and methodological suggestions mentioned, the empirical studies have gradually varied following those indications.

From the GLCS a greater attention to the relationship between Integrated Reporting and ICD emerged. Finally, the VOSclustering analysis identified five clusters of keywords that represent the following topics:

1. Features of empirical research in ICD field (sample, data sources, contents examined);
2. ICD determinants;
3. IC management to get disclosure information;
4. The three categories of IC;
5. The state of the art (with additional contextual variables). Moreover, heterogeneity of the context is clear in almost all the clusters but helps defining which are the main environments in which ICD research has been conducted; in terms of geographical areas, e.g. Portugal, the UK and Singapore are the main countries investigated, while Listed Companies and Banks are the types of organizations mainly selected for the sample. Among the most recent keywords used—in addition to IC, ICD, voluntary disclosure and listed companies—"*Structural Capital*" and "*Relational Capital*" are also two of the most cited. This result confirms that the approach towards ICD research is changing: authors are not focused solely on a unique dimension of IC anymore; instead, the tendency of splitting IC into its three categories for research purposes is proved and compliant to the need of valuating the ICD quality as emerged throughout the previous analysis. In conclusion, Kleinberg's algorithm showed the frequency and timespan of each keyword used, highlighting com-

pletely different trends in the first and last ten years of analysis (2002–2012 and 2012–2022). It confirmed the results of previous analysis from another perspective. The main bursts concern:

1. The definition of intangible assets.
2. The use of Annual Reports for the content analysis in less recent papers, from 2004 until 2012, before the methodological critique.
3. Some specific contexts, determinants and tools, e.g. Integrated Reporting being a very innovative tool that discloses information on the three categories of IC, was the most used in the newest papers corroborating the current trends just mentioned. The most recent topic regards the governance variables which interact with ICD and the IC performance. Other relevant issues concern the ICD of the universities and the interest in constructing a clear and common theoretical framework, two themes that already emerged through the Main Path analysis.

Intellectual Capital Disclosure (ICD) refers to the reporting and disclosure of intangible assets or intellectual capital by companies in their financial statements or other public documents. Intellectual capital may include assets such as patents, trademarks, copyrights, brand value, Human Capital (such as employee skills and knowledge) and Organizational Capital (such as processes and systems). Disclosure of intellectual capital provides stakeholders with information about the value and importance of these intangible assets to the company's overall performance and future prospects.

In the context of sustainability reporting, intellectual capital disclosure may become increasingly important.

In fact, regulatory frameworks and reporting standards, such as the Global Reporting Initiative (GRI) require companies to disclose information on intangible assets, including intellectual capital, as part of their sustainability reporting obligations.

Integrating intellectual capital disclosure into sustainability reporting can enhance transparency, improve risk management, demonstrate value creation efforts and ensure compliance with evolving reporting standards and stakeholder expectations.

The analysis provides an answer also to the second research question about the future of ICD research. In particular, six main themes are identified, i.e.:

1. "Data source";
2. "Reporting vs Disclosure";
3. "Static vs dynamic studies";
4. "Comparative studies",
5. "IC as a whole";
6. "Methodology".

In the first place, those categories make it very clear that the biggest improvements should be performed in the methodological approach. Regarding the sources to get information, findings reveal that future empirical studies on ICD should use new data sources (e.g. website, Integrated Reporting) and apply new theories (Beattie & Thomson, 2007; Dumay & Cai, 2014, 2015; Garanina & Dumay, 2017; Ndou et al., 2018; Dumay et al., 2019; Beretta et al., 2019; Salvi et al., 2020a, 2020b; Aversano et al., 2020; Brusca et al., 2020; De Villiers & Sharma, 2020; Vitolla et al., 2020; Musleh Al-Sartawi, 2021). In addition, researchers should consider companies' involuntary disclosure and internal IC practices more, giving more space to interviews, field notes, tweets, blogs and news media for conducting a day-by-day analysis and shift their attention from mere reporting tools to a broader spectrum of disclosure means (Dumay & Cai, 2014, 2015; Dumay, 2016; Garanina & Dumay, 2017). Regarding the horizon of analysis, future studies should be longitudinal and comparative (Brennan, 2001; Bozzolan et al., 2003; Bukh et al., 2005 Guthrie et al., 2006; Goh & Lim, 2004; Li et al., 2008; Cerbioni & Parbonetti, 2007; Brusca et al., 2020; Aversano et al., 2020; Nicolò et al., 2021) to observe the evolution of the ICD practices and the impact of some contextual features. To gather effective updates and compare previous research findings, authors should also consider new variables that could have an impact on ICD, such as the corporate culture (Goh & Lim, 2004). They should also avoid considering IC as an unit but study the three IC categories (Organizational, Human and Relational Capital) separately in order to obtain deeper insights (Bozzolan et al., 2003; Goh & Lim, 2004; Vergauwen et al., 2007; Cinquini et al., 2012; Aversano et al., 2020). Finally, from a strictly methodological and theoretical point of view, they should mix different methodologies (such as content analysis and questionnaire, interviews, etc.) to triangularize the data inputs and obtain more robust findings. Moreover, they should try to elaborate a shared framework to investigate this topic (Guthrie et al., 2004; Beattie &

Thomson, 2007; Dumay & Cai, 2014, 2015), which can be used to obtain comparable knowledge through different contexts over time.

This chapter provides four main contributions. Firstly, it helps to develop new knowledge in the accounting field through new and innovative methodologies (Dumay & Cai, 2014, 2015; Dumay, 2016; Massaro et al., 2016). Secondly, it applies the SLNA for performing the analysis of the literature, a technique that helps in reducing the level of subjectivity, thanks to the presence of a rigid protocol and rules (Dumay & Cai, 2014, 2015; Dumay, 2014, 2016; Massaro et al., 2016). Finally, it maps the current knowledge in the ICD field, depicts its evolution over time, highlights the most important issues debated by researchers over time, and proposes both future research topics and methodological suggestions according to the previous critique. Finally, findings confirm that the methodology adopted by this chapter is suitable to carry out a structured literature review in the field of non-financial reporting, in compliance with the findings of Afeltra et al. (2022), and that the SLNA is a useful research tool to support dynamic analyses and develop agendas for future research in the accounting field, as suggested also by Tettamanzi and Comerio (2019).

There are also some limitations to the methodology applied. First, the citation data are entirely extracted from the Scopus database and represent, for this reason, only a part of the total amount of the scientific publications. It is, however, the most appropriate database for performing an SLNA because it is the largest abstract and citation database. Second, the so-called Matthew effect (researchers tend to cite papers which have already received a high number of citations, since they are considered more reliable as a source of information) is not neutralized. This limitation is partially solved by the fact that the findings of this study are not only based on the number of citations, but they are integrated with the keywords' analysis. Third, the process of data selection, the analysis and the interpretation of the findings are not completely free of subjectivity problems, even if the level is lower with respect to other methodologies.

References

Abeysekera, I. (2006). The project of ICD: Researching the research. *Journal of Intellectual Capital, 7*(1), 61–77.

Abeysekera, I., & Guthrie, J. (2003). An empirical investigation of annual reporting trends of intellectual capital in Sri Lanka. *Critical Perspectives on Accounting, 16*(3), 151–163.

Abeysekera, L., & Guthrie, J. (2004). Human capital reporting in a developing nation. *British Accounting Review, 36*(3), 251–268.

Abhayawansa, S., & Guthrie, J. (2014). Importance of intellectual capital information: A study of Australian analyst reports. *Australian Accounting Review, 68*(24), 66–83.

Afeltra, G., Alerasoul, A., & Usman, B. (2022). Board of Directors and Corporate Social Reporting: A systematic literature network analysis. *Accounting in Europe, 19*(1), 48–77.

Amorelli, M.-F., & García-Sánchez, I.-M. (2020). Critical mass of female directors, human capital, and stakeholder engagement by corporate social reporting. *Corporate Social Responsibility and Environmental Management, 27*(1), 204–221.

An, Y., Sharma, U., & Wang, Z. (2015). Towards a conceptual template for intellectual capital measurement and reporting. *International Journal of Business and Management, 10*(7), 236–245.

Aversano, N., Nicolò, G., Sannino, G., & Polcini, P. T. (2020). The integrated plan in Italian public universities: New patterns in intellectual capital disclosure. *Meditari Accountancy Research, 28*(4), 655–679.

Barile, S., Saviano, M., Polese, F., & Caputo, F. (2015). T-shaped people for addressing the global challenge of sustainability. In E. Gummesson, C. Mele, & F. Polese (Eds.), *Service dominant logic, network and systems theory and service science: Integrating three perspectives for a new service agenda* (pp. 1–24). Giannini.

Beattie, V., & Thomson, S. J. (2007). Lifting the lid on the use of content analysis to investigate intellectual capital disclosure. *Accounting Forum, 31*(2), 129–163.

Beretta, V., Demartini, C., & Trucco, S. (2019). Does environmental, social and governance performance influence intellectual capital disclosure tone in integrated reporting? *Journal of Intellectual Capital, 20*(1), 100–124.

Bozzolan, S., Favotto, F., & Ricceri, F. (2003). Italian annual intellectual capital disclosure, an empirical analysis. *Journal of Intellectual Capital, 4*(4), 543–558.

Brennan, N. (2001). Reporting intellectual capital in annual reports: Evidence from Ireland. *Accounting, Auditing & Accountability Journal, 14*(4), 423–436.

Brusca, I., Cohen, S., Manes-Rossi, F., & Nicolò, G. (2020). Intellectual capital disclosure and academic rankings in European universities. *Meditari Accountancy Research, 28*(1), 51–71.

Bukh, P. N., Larsen, H. T., & Mouritsen, J. (2001). Constructing intellectual capital statements. *Scandinavian Journal of Management, 17*(1), 87–108.

Bukh, P. N., Nielsen, C. H., Gormsen, P. U., & Mouritsen, J. (2005). Disclosure of intellectual capital indicators in Danish IPO prospectuses. *Accounting Auditing and Accountability Journal, 18*(6), 731–732.

Burgman, R., & Roos, G. (2006). *Operational and intellectual capital reporting: Risk-based decision models for identifying, quantifying and reporting on enterprise value drivers in a 'managing for value' framework*. Paper presented at the

2nd EIASM Workshop on Visualising, Measuring and Managing Intellectual Capital and Intangibles, Maastricht, 25–27 October.

Burgman, R., & Roos, G. (2007). The importance of intellectual capital reporting: Evidence and implications. *Journal of Intellectual Capital, 8*(1), 7–51.

Campanella, F., Della Peruta, M. R., & Del Giudice, M. (2013). The role of sociocultural background on the characteristics and the financing of youth entrepreneurship: An exploratory study of university graduates in Italy. *Journal of the Knowledge Economy, 4*(3), 244–259.

Carter, C. R., & Easton, P. L. (2011). Sustainable supply chain management: Evolution and future directions. *International Journal of Physical Distribution and Logistics Management, 41*(1), 46–62.

Castilla-Polo, F., & Gallardo-Vázquez, D. (2016). The main topics of research on disclosures of intangible assets. A critical review. *Accounting, Auditing & Accountability Journal, 29*(2), 323–356.

Castilla-Polo, F., & Ruiz-Rodríguez, C. (2017). Content analysis within intangible assets disclosure. A structured literature review. *Journal of Intellectual Capital, 18*(3), 506–543.

Cerbioni, F., & Parbonetti, A. (2007). Exploring the effects of corporate governance on intellectual capital disclosure: An analysis of European biotechnology companies. *European Accounting Review, 16*(4), 791–826.

Chen, R., & Sharma, S. K. (2013). Self-disclosure at social networking sites: An exploration through relational capitals. *Information Systems Frontiers, 15*(2), 269–278.

Chiucchi, M. S., & Dumay, J. (2015). Unlocking intellectual capital. *Journal of Intellectual Capital, 16*(2), 305–330.

Cinquini, L., Passetti, E., Tenucci, A., & Frey, M. (2012). Analyzing intellectual capital information in sustainability reports: Some empirical evidence. *Journal of Intellectual Capital, 13*(4), 531–561.

Colicchia, C., & Strozzi, F. (2012). Supply chain risk management: A new methodology for a systematic literature review. *Supply Chain Management: An International Journal, 17*(4), 403–418.

Cuozzo, B., Dumay, J., Palmaccio, M., & Lombardi, R. (2017). Intellectual capital disclosure: a structured literature review. *Journal of Intellectual Capital, 18*(1), 9–28.

Dawson, S., Gasević, D., Siemens, G., & Joksimovic, S. (2014). Current state and future trends: A citation network analysis of the learning analytics field. In *Proceedings of the fourth international conference on learning analytics and knowledge* (pp. 231–240).

De Nooy, W., Mrvar, A., & Batagelj, V. (2011). *Exploratory social network analysis with Pajek: revised and expanded second edition, structural analysis in the social sciences*. Cambridge University Press.

Denyer, D., & Tranfield, D. (2009). Producing a systematic review. In D. Buchanan & A. Bryman (Eds.), *The sage handbook of organizational research methods* (pp. 671–689). Sage.

De Villiers, C., & Sharma, U. (2020). A critical reflection on the future of financial, intellectual capital, sustainability and integrated reporting. *Critical Perspectives on Accounting, 70,* 101999.

Ding, Y., Chowdhury, G. G., & Foo, S. (2001). Bibliometric cartography of information retrieval research by using co-word analysis. *Information Processing & Management, 37*(6), 817–842.

Dumay, J. (2014). 15 years of the journal of intellectual capital and counting: A manifesto for transformational IC research. *Journal of Intellectual Capital, 15*(1), 2–37.

Dumay, J. (2016). A critical reflection on the future of intellectual capital: From reporting to disclosure. *Journal of Intellectual Capital, 17*(1), 168–184.

Dumay, J., & Cai, L. (2014). A review and critique of content analysis as a methodology for inquiring into IC disclosure. *Journal of Intellectual Capital, 15*(2), 264–290.

Dumay, J., & Cai, L. (2015). Using content analysis as a research methodology for investigating intellectual capital disclosure: A critique. *Journal of Intellectual Capital, 16*(1), 1–36.

Dumay, J., La Torre, M., & Farneti, F. (2019). Developing trust through stewardship: Implications for intellectual capital, integrated reporting, and the EU directive 2014/95/EU. *Journal of Intellectual Capital, 20*(1), 11–39.

Eccles, R. G., & Mavrinac, S. C. (1995). Improving the corporate disclosure process. *MIT Sloan Management Review, 36*(4), 11–25.

Edvinsson, L. (2013). IC 21: Reflections from 21 years of IC practice and theory. *Journal of Intellectual Capital, 14*(1), 163–172.

Enache, L., & Srivastava, A. (2018). Should intangible investments be reported separately or commingled with operating expenses? New evidence. *Management Science, 64*(7), 3446–3468.

Garanina, T., & Dumay, J. (2017). Forward-looking intellectual capital disclosure in IPOs. *Journal of Intellectual Capital, 18*(1), 128–148.

Goh, P. C., & Lim, K. P. (2004). Disclosing intellectual capital in company annual reports: Evidence from Malaysia. *Journal of Intellectual Capital, 5*(3), 500–510.

Gray, D., Rastas, T., & Roos, G. (2004). What intangible resources do companies value, measure and report? A synthesis of UK and Finnish research. *International Journal of Learning and Intellectual Capital, 1*(3), 242–261.

Greenhalgh, C., & Longland, M. (2005). Running to stand still? The value of R&D, patents and trade marks in innovating manufacturing firms. *International Journal of the Economics of Business, 12*(3), 307–328.

Guthrie, J., Petty, R., Wells, R. (1999). *There is no accounting for intellectual capital in Australia: A review of annual reporting practices and the internal mea-*

surement of intangibles. Paper presented at OECD Symposium on Measuring and Reporting of Intellectual Capital, Amsterdam.

Guthrie, J., Petty, R., & Ricceri, F. (2006). The voluntary reporting of intellectual capital: Comparing evidence from Hong Kong and Australia. *Journal of Intellectual Capital, 7*(2), 254–271.

Guthrie, J., Petty, R., Yongvanich, K., & Ricceri, F. (2004). Using content analysis as a research method to inquire into intellectual capital reporting. *Journal of Intellectual Capital, 5*(2), 282–293.

Hsu, Y., & Fang, W. (2009). Intellectual capital and new product development performance: The mediating role of organizational learning capability. *Technological Forecasting and Social Change, 76*(5), 664–677.

Hyvonen, S., & Tuominen, M. (2006). Entrepreneurial innovations, market-driven intangibles and learning orientation: Critical indicators for performance advantages in SMEs. *International Journal of Management and Decision Making, 7*(6), 643–660.

Integrated Reporting Framework (IRF). (2021). Document available at: https://www.integratedreporting.org/resource/international-ir-framework/

Kim, S., Colicchia, C., & Menachof, D. (2018). Ethical sourcing: An analysis of the literature and implications for future research. *Journal of Business Ethics, 152*(4), 1033–1052.

Kleinberg, J. (2003). Burst and hierarchical structure in streams. *Data Mining and Knowledge Discovery, 7*(4), 373–397.

Knoke, D., & Yang, S. (2008). *Social network analysis* (Vol. 154).

Kong, E. (2010). Intellectual capital management enablers: A structural equation modeling analysis. *Journal of Business Ethics, 93*(3), 373–391.

Kwee, K. C. (2008). Intellectual capital: Definitions, categorization and reporting models. *Journal of Intellectual Capital, 9*(4), 609–638.

Lev, B. (2001). *Intangibles management, measurement, and reporting*. Brookings Institution Press.

Li, J., Pike, R., & Haniffa, R. (2008). Intellectual capital disclosure and corporate governance structure in UK firms. *Accounting Business Research, 38*(2), 137–159.

Li, S. T., Tsa, M. H., & Lin, C. (2010). Building a taxonomy of a firm's knowledge assets: A perspective of durability and profitability. *Journal of Information Science, 36*(1), 36–56.

Lin, C. S., & Huang, C. P. (2011). Measuring competitive advantage with an asset-light valuation model. *African Journal of Business Management, 5*(1), 5100–5108.

Lucio-Arias, D., & Leydesdorff, L. (2008). Main-path analysis and path-dependent transitions in HistCite™-based historiograms. *Journal of the Association for Information Science and Technology, 59*(12), 1948–1962.

Martín, G., Delgado, M., López, P., & Navas, J. E. (2011). Towards 'an intellectual capital-based view of the firm': Origins and nature. *Journal of Business Ethics, 98*(4), 649–662.

Martín-de Castro, G., Díez-Vial, I., & Delgado-Verde, M. (2019). Intellectual capital and the firm: Evolution and research trends. *Journal of Intellectual Capital, 20*(4), 555–580.

Massaro, M., Dumay, J., & Guthrie, J. (2016). On the shoulders of giants: Undertaking a structured literature review in accounting. *Accounting, Auditing & Accountability Journal, 29*(5), 767–801.

Musleh Al-Sartawi, A. M. A. (2021). Social media disclosure of intellectual capital and firm value. *International Journal of Learning and Intellectual Capital, 17*(4), 312–323.

Nadeem, M. (2019). Does board gender diversity influence voluntary disclosure of intellectual capital in initial public offering prospectuses? Evidence from China. *Corporate Governance: An International Review, 28*(2), 100–118.

Ndou, V., Secundo, G., Dumay, J., & Gjevori, E. (2018). Understanding intellectual capital disclosure in online media Big Data. *Meditari Accountancy Research, 26*(3), 499–530.

Nicolò, G., Aversano, N., Sannino, G., & Tartaglia Polcini, P. (2021). ICD corporate communication and its determinants: Evidence from Italian listed companies' websites. *Meditari Accountancy Research, 29*(5), 1209–1232.

Noruzi, A. (2005). Google scholar: The new generation of citation indexes. *International Journal of Libraries and Information Services Libri, 55*(4), 170–180.

Oliveira, L., Lima Rodrigues, L., & Craig, R. (2010). Intellectual capital reporting in sustainability reports. *Journal of Intellectual Capital, 11*(4), 575–594.

Pablos, P. O. D. (2002). Evidence of intellectual capital measurement from Asia, Europe and the Middle East. *Journal of Intellectual Capital, 3*(3), 287–302.

Petticrew, M., & Roberts, H. (2008). *Systematic reviews in the social sciences: A practical guide* (Kindle ed.). Wiley-Blackwell.

Rashman, L., Withers, E., & Hartley, J. (2009). Organizational learning and knowledge in public service organizations: A systematic review of the literature. *International Journal of Management Reviews, 11*(4), 463–494.

Roulstone, D. T. (2011). Discussion of intangible investment and the importance of firm-specific factors in the determination of earnings. *Review of Accounting Studies, 16*(3), 574–586.

Salvi, A., Vitolla, F., Giakoumelou, A., Raimo, N., & Rubino, M. (2020a). Intellectual capital disclosure in integrated reports: The effect on firm value. *Technological Forecasting and Social Change, 160*, 120228.

Salvi, A., Vitolla, F., Raimo, N., Rubino, M., & Petruzzella, F. (2020b). Does intellectual capital disclosure affect the cost of equity capital? An empirical analysis in the integrated reporting context. *Journal of Intellectual Capital,* 21(6), 985–1007.

Sällebrant, T., Hansen, J., Bontis, N., & Hofman-Bang, P. (2007). Managing risk with intellectual capital statements. *Management Decision,* 45(9), 1470–1483.

Schneider, A., & Samkin, G. (2008). Intellectual capital reporting by the New Zealand local government sector. *Journal of Intellectual Capital,* 9(3), 456–486.

Sharma, R. S., Hui, P. T. Y., & Tan, M. W. (2007). Value-added knowledge management for financial performance: The case of an East Asian conglomerate. *Vine: The Journal of Information and Knowledge Management Systems,* 37(4), 484–501.

Singh, I., & Van der Zahn, J.-L. W. M. (2008). Determinants of intellectual capital disclosure in prospectuses of initial public offerings. *Accounting and Business Research,* 38(5), 409–431.

Singh, R. D., & Narwal, K. P. (2015). Intellectual capital and its consequences on company performance: A study of Indian sectors. *International Journal of Learning and Intellectual Capital,* 12(3), 300–322.

Skoog, M. (2003). Visualizing value creation through the management control of intangibles. *Journal of Intellectual Capital,* 4(4), 487–504.

Stewart, T. A. (1997). *Intellectual capital: The new wealth of organizations.* Doubleday Dell Publishing Group.

Strozzi, F., Colicchia, C., Creazza, A., & Noè, C. (2017). Literature review on the 'Smart Factory' concept using bibliometric tools. *International Journal of Production Research,* 55(22), 6572–6591.

Sveiby, K. E. (1997). *The new organizational wealth: Managing and measuring knowledge-based assets.* Berrett-Koehler.

Tettamanzi, P., & Comerio, N. (2019). Systematic literature network analysis in accounting: A first application on integrated reporting research. *Financial reporting: Bilancio, controlli e comunicazione d'azienda,* 2, 73–95.

Vandemaele, S. N., Vergauwen, P. G. M. C., & Smits, A. J. (2005). Intellectual capital disclosure in The Netherlands, Sweden and the UK, a longitudinal and comparative study. *Journal of Intellectual Capital,* 6(3), 417–426.

Van Eck, N. J., & Waltman, L. (2010). Software survey: VOSviewer, a computer program for bibliometric mapping. *Scientometrics,* 84(2), 523–538.

Vergauwen, P., Bollen, L., & Oirbans, E. (2007). Intellectual capital disclosure and intangible value drivers: An empirical study. *Management Decision,* 45(7), 1163–1180.

Vitolla, F., Raimo, N., Marrone, A., & Rubino, M. (2020). The role of board of directors in intellectual capital disclosure after the advent of integrated reporting. *Corporate Social Responsibility and Environmental Management,* 27(5), 2188–2200.

Waltman, L., Van Eck, N. J., & Noyons, E. C. (2010). A unified approach to mapping and clustering of bibliometric networks. *Journal of Informetrics*, 4(4), 629–635.

Williams, S. M. (2001). Is intellectual capital performance and disclosure practises related? *Journal of Intellectual Capital*, 2(3), 192–203.

Wong, M., & Gardner, C. (2005). *Intellectual capital disclosure: New Zealand evidence*. Paper presented at the AFFANZ 2005 Conference, Melbourne.

Yi, A., & Davey, H. (2010). Intellectual capital disclosure in Chinese (mainland) companies. *Journal of Intellectual Capital*, 11(3), 326–347.

Zhao, D., & Strotmann, A. (2015). Analysis and visualization of citation networks. *Synthesis Lectures on Information Concepts, Retrieval, and Services*, 7(1), 1–207.

CHAPTER 2

The Circular Economy Disclosure in EU Setting

*Isabel-María García-Sánchez and
Saudi-Yulieth Enciso-Alfaro*

Abstract Social demands for the conservation and care of ecosystems have led to the approach of a circular economy (CE) supported by clean technologies toward which companies must converge. Making it is necessary to pay special attention to planetary boundaries with respect to which interest groups demand greater disclosure of information for a critical assessment of its transition. In this research, we present an overview of the resource use and circular economy information disclosure by companies in the European context. In order to achieve our objective, we employed a sample of 1315 companies over the period 2014–2022. Although we have

---

I.-M. García-Sánchez (✉) • S.-Y. Enciso-Alfaro
University of Salamanca, Salamanca, Spain
e-mail: lajefa@usal.es; idu019523@usal.es

identified a significant commitment, in order to align with the current EU requirements, it would be beneficial for the companies to provide more detailed information.

Keywords Resource use • Circular economy • Information disclosure • European Union

2.1 Introduction

According to European Parliament (2024), the scarcity of resources and the environmental risks derived from inappropriate use of nature require moving from an economic model of "take, make and dispose" to an economy focused on the care of ecosystems, which is committed to emissions neutrality, toxic-free and completely circular by 2050.

In this way, the circular economy (CE, hereinafter) is proposed as a regenerative system that seeks to optimize the consumption of resources (renewable and non-renewable) and raw materials (virgin and non-virgin), as well as the reduction of waste generation and polluting emissions to ecosystems, and energy waste, through the repair, reuse, reconditioning and recycling of goods, materials and other resources (Geissdoerfer et al., 2017; Korhonen et al., 2018; Kirchherr et al., 2023). Proposals also recognized and addressed in Delegated Regulation (EU) 2023/2772 of the European Parliament (2023), specifically in European sustainability reporting standards (ESRS) E5, relating to the disclosure of business information on resource use and CE.

Additionally, according to Desing et al. (2020) and Enciso-Alfaro and García-Sánchez (2024), to maintain the resource base and environmental conditions necessary for human existence on the Earth planet, the CE as an economic paradigm must respect the terrestrial capacity, materialized through natural processes related to: climate change, biosphere integrity, land-system change, freshwater change, biogeochemical flows, ocean acidification, atmospheric aerosol loading, stratospheric ozone depletion and novel entities, according to the Planetary Boundaries framework developed by Richardson et al. (2023), Rockström et al. (2009), among others.

The Planetary Boundaries framework establishes a "safe space" within which all human activity must take place (Rockström et al., 2009; Steffen et al., 2015), establishing variables to monitor the current state of each process and identify whether they are within limits that facilitate or threaten the ecosystem life and the humankind life.

For example, for the process called *freshwater change*, the existence of drivers of freshwater depletion is established, such as excessive use in irrigation activities, which in turn have impacts on the moisture levels of the surrounding soil (Porkka et al., 2024). In this sense, both river flow (blue water) and soil humidity (green water) are key elements in the water cycle, so establishing optimal maintenance levels provides a guide or "safe space" with which companies can guide the development of their activities without affecting or transgressing the desirable limit values.[1]

In this way, Enciso-Alfaro and García-Sánchez (2024) stated that different initiatives reported by companies and related to CE, such as the recycling of water in production processes or the establishment of gray water systems in buildings, can contribute to maintaining an optimal level of natural sources by reducing their consumption, it being important to report detailed information that allows, from a business approach, to evaluate the business transition toward a sustainable CE that contributes to the stability and resilience of natural processes in accordance with the Planetary Boundaries framework.

On the other hand, previous literature related to the disclosure of information on environmental sustainability continues to demand the development of studies that expand the understanding of the effects of the enactment of laws and/or agreements that seek greater reporting of business activities carried out against the demands to care for and protect natural resources and the commitment to circular and sustainable business models, due to the fact that some results recorded in previous literature are not conclusive as they present positive and/or non-significant results of the institutional pressure associated with environments with strong regulation.

The findings of García-Sánchez et al. (2023), Liu and Guo (2023) and Mio et al. (2020) indicate a favorable effect of the establishment of environmental regulation on the disclosure of environmental sustainability information, in terms of quantity and quality. Furthermore, Enciso-Alfaro and García-Sánchez (2024) point out the commitment of companies to report their progress in terms of transition toward a CE business model.

However, other research such as those by Cubilla-Montilla et al. (2020) and Wukich et al. (2023) indicates that the disclosure of information on environmental sustainability by companies is influenced in a non-significant

---

[1] To learn more about the planetary limits, as well as the ranges in which each of them must operate to maintain adequate environmental conditions for human and natural life, you can consult: https://www.stockholmresilience.org/publications.html

way by coercive pressures, with cultural and mimetic patterns being those that determine changes in the levels of disclosure of corporate information.

In this sense, the objective of this work is to present an overview of the disclosure of information regarding the use of resources and CE by companies in the European context through the design of the indicator, ScoreIRUCE, which covers the goals and initiatives to the use of resources and CE, as well as the entry and exit of resources, and the forecast of future risk and opportunities.

To do this, in this chapter we analyze the information reported by companies located in member countries of the European Union (EU) on the aforementioned topics. In this regard, our sample is composed of 1315 companies for the period 2014–2022, constituting an unbalanced data panel of 7909 observations. These companies operate in 23 industries and are located in 20 European countries.

The consideration of the European scope is due, first of all, to the relevance and economic impact of the EU in the world economy. According to data reported by World Bank (2024), Europe occupies third place in the world economy, with an approximate GDP of 27.2 million dollars by 2023. Secondly, the important commitment of the EU to place to the European Union on the path of restoring and recovering biodiversity within the next six years. According to European Parliament (2020), the business community plays a very relevant role in this sense. Thus, business strategies and the consequent disclosure of information are a source of valuable information to evaluate progress and promote policies that help companies align their operations with ecosystem well-being.

The period of analysis has been characterized by greater international recognition and awareness of ecosystem conservation through agreements such as those of Paris (United Nations, 2015) or the European Green Deal (European Commission, 2021). Likewise, unexpected events have occurred that could have affected the business' ability to establish environmentally appropriate strategies and the consequent disclosure of information, such as the COVID-19 pandemic and the war in Ukraine.

Our results show that European companies are committed to the disclosure of information related to the optimization of the use of resources and the CE, disclosing information for 45% of the items analyzed in accordance with the requirements that Directive 2022/2464/EU imposes currently. In this regard, it should be indicated that the requirements were not in force in the period analyzed. However, this figure indicates that companies have to continue developing initiatives that allow their operations to align with environmental demands and conservation of European natural wealth.

Thus, we contribute to previous literature by designing a score that allows us to assess the degree of business progress toward an economy integrated with land capacity, through the reporting of corporate information related to the use of resources and a solid transition toward a CE. Providing an overview that allows us to know the aspects in which European companies would need greater institutional support to achieve full implementation, such as the installation of technology for the recycling of water resources.

2.2 European Institutional Forces

Institutional Theory

The institutional theory (DiMaggio & Powell, 1983) establishes that different factors or institutional pressures cause companies to seek to align themselves with formal or informal demands to guarantee their survival, leading them to disclose information about certain aspects of the development of their activities, such as related to environmental sustainability, entering a process called isomorphism.

In this sense, DiMaggio and Powell (1983) indicate that isomorphism can occur due to three forces: coercive, mimetic and/or normative. Coercive force comes from the establishment of laws or rules (García-Sánchez et al., 2022). The mimetic force is associated with the practices of the most relevant companies in a sector (Zampone et al., 2022), and the normative force arises from professional norms and values (Martínez-Ferrero & García-Sánchez, 2017).

In this work, we focus on coercive and normative isomorphisms since the disclosure of information on resource use and CE is subject to national and supranational agency guidelines, as well as professional expectations.

European Pronouncements on Sustainability Information Disclosure

Within the EU there are various pronouncements on the disclosure of non-financial information or on sustainability matters. In this regard, Directive (EU) 2014/95/EU of the European Parliament (NFR Directive) stands out for being the pioneer. This Directive (EU) 2014/95 represented a significant advance in corporate transparency by imposing on large companies and public interest groups the obligation to disclose non-financial and diversity information, promoting transparency in environmental, social and governance aspects. It established the obligation that companies located in EU member countries must report on aspects that arise from the development of business activities and that in turn are or may condition correct environmental functioning (i.e., polluting emissions).

Furthermore, with the purpose of guiding companies in the specific aspects to disclose in terms of non-financial information, in 2017, the European Parliament published Directive C/2017/4234, which details the fundamental principles for the preparation and the content of non-financial statements, especially so that environmental information is relevant, coherent, useful and of high quality. Later, in 2019, this same institution published a supplement for the presentation of climate-related information, the non-binding guideline C/2019/4490, whose purpose is to increase the quality and quantity of the data reported.

Additionally, with the purpose of monitoring progress toward an economy dissociated from the use of resources, the European Parliament approved Directive (EU) 2022/2464 of the European Parliament and of the Council, of December 14, 2022, on corporate information in matters of Sustainability Directive (CSRD Directive), establishing a legal framework that guarantees that sustainability information disclosed by companies meets the information needs of different users. Regarding the topic addressed in this chapter, it requires companies to report on the goals, initiatives and results achieved in optimizing the use of resources in production processes (input and output of resources) and CE.

In this way, the previous directives and guidelines configure a set of coercive and regulatory pressures that demand a prompt business response that satisfies the need for clear and detailed information on the protection, conservation and restoration of the natural capital of the European Union.

2.3 Sample Configuration and Presentation of the Score for Dissemination of Information on Resource Use and Circular Economy

Sample Selection

To achieve our research objective, we select the most relevant companies located in European countries given their environmental and social impact, and the ability to promote sustainable actions due to their volume of resources and capabilities. The database used to identify companies' sustainability information is Refinitiv.

In accordance with previous literature (i.e., Arvidsson & Dumay, 2022; Monteiro et al., 2021; Aureli et al., 2020), sustainability reports include detailed information on the use of renewable and non-renewable natural resources, virgin raw materials and non-virgin, as well as the strategies and initiatives developed by companies to achieve the circular business transition. This information is tabulated in the indicated database. In this sense, the sample selection criterion used corresponds to the availability of data on sustainability. Its application has determined a sample made up of 7909 observations corresponding to an unbalanced data panel of 1315 companies for the period 2014–2022. These European companies develop their economic activity in 23 industries and their headquarters are located in 20 different European countries.

Based on the information available on environmental matters, Table 2.1 presents the ScoreIRUCE, which we have designed by aggregating dichotomous variables that identify information related to the use of resources and EC, taking values between 0 and 23.

Design of the Score of Information on the Use of Resources and CE

In order to determine the material information that companies must report on the use of resources and CE, proposals from international organizations were considered, i.e., GRI—Global Reporting Initiative, SASB—Sustainability Accounting Standard Board and especially as specified in the European sustainability reporting standards (ESRS) E5 Resource use and circular economy of Commission Delegated Regulation (EU) 2023/2772 of July 31, 2023, supplementing Directive 2013/34/EU of the European Parliament and of the Council as regards sustainability reporting standards.

In this sense, we configure the ScoreIRUCE by adding a set of informative items that allow the identification of (i) the goals to be achieved in terms of resource use and CE through the TDOs sub-score that has seven items; (ii) initiatives for the use of resources and CE, through the INITs sub-score with eight items; (iii) the entry of resources, through the RCMs sub-score that has five items; (iv) the output of resources and/or recycling of waste, through the RFLs sub-score that brings together four items as well as (v) the forecast of future risks and opportunities in terms of environmental proactivity through the FHP item. Table 2.1 details the items that make up each sub-score and, by aggregation, the global disclosure score.

Table 2.1 ScoreIRUCE: composition

	Item	Description	Relative frequency
Goals pertaining to the utilization of resources and the CE	TDO1	Proposed targets for reducing the use of natural resources (renewable and non-renewable)	89.3%
	TDO2	Natural resource and material efficiency targets (virgin and non-virgin)	38.1%
	TDO3	Proposed water resource efficiency targets	17.5%
	TDO4	Proposed energy efficiency targets	33.7%
	TDO5	Proposed targets for the exclusive use of sustainable packaging	22.3%
	TDO6	Proposed targets for reducing the overall ecosystem impact of the supply chain through joint and collaborative working	72.2%
	TDO7	Proposed targets for reducing, recycling, reusing, replacing, treating or phasing out all waste, including electronic waste	78.7%
	Sub-score TDOs (0–7 points)		50.3% (mean points = 3.52)

(*continued*)

Table 2.1 (continued)

	Item	Description	Relative frequency
Initiatives pertaining to the utilization of resources and the CE	INIT1	Initiatives to assess the life cycle of materials for optimization	49.1%
	INIT2	Initiatives taken to reduce, reuse, substitute or eliminate chemical/toxic substances (i.e., sox: Sulfur oxides; NOx: Nitrogen oxides, volatile organic compounds [VOC] or particulate matter less than ten microns in diameter [PM10])	31.3%
	INIT3	Initiatives taken to retrofit buildings according to eco-efficiency criteria	27.5%
	INIT4	Initiatives taken to reduce the impact of natural resource use on biodiversity	30.1%
	INIT5	Initiatives or partnerships established with other organizations (e.g., NGOs, governments, etc.) to restore ecosystem damage	47.1%
	INIT6	Initiatives established to develop clean and renewable energy technologies for inter-organizational use (i.e., wind, geothermal, biomass, etc.) to replace fossil fuels	19.7%
	INIT7	Initiatives set up to develop technologies for the treatment or purification of water or to improve the efficiency of the use of water resources	3.7%
	INIT8	Initiatives set up to develop products that can have a positive impact on the environment at the end of their optimal useful life (e.g., 100% biodegradable packaging, seed labels, etc.)	61.4%
	Sub-score INITs (0–8 points)		33.7% (mean points = 2.70)

(*continued*)

Table 2.1 (continued)

	Item	Description	Relative frequency
Resources inflow	RCM1	Energy consumption from non-renewable sources in gigajoules	73.0%
	RCM2	Energy consumption from renewable sources in gigajoules	39.7%
	RCM3	Water consumption from natural sources in cubic meters	61.2%
	RCM4	Water consumption from recycled sources in cubic meters	8.4%
	RCM5	Consumption of recycled materials	1.6%
	Sub-score RCMs (0–5 points)		36.8% (mean points = 1.84)
Resources outflow	RFL1	Total non-hazardous waste generated in tonnes	58.3%
	RFL2	Total amount of hazardous waste generated in tonnes	47.6%
	RFL3	Total amount of waste recycled or reused in tonnes	47.6%
	RFL4	Total tons of recycled plastic	47.4%
	Sub-score RFLs (0–4 points)		50.2% (mean points = 2.01)
Forecasts, hazards and prospects	FHP	Proactive environmental investments or provisions are measures taken to reduce future risks or increase future opportunities	33.2%
	FHP sub-score (0–1 point)		33.2% (mean point = 0.33)
ScoreIRUCE (0–25 points)			45.20% (mean points = 10.40)

Source: Own elaboration

Preliminary Results

According to the statistical information reflected in Table 2.1, in relation to the information that companies report on the goals set by the companies, the one related to the reduction of the use of natural resources (renewable and non-renewable) stands out, being reported by 89.3% of the European companies analyzed. Next, it can be seen that 78.7% of the companies analyzed report on the proposed goals to reduce, recycle, reuse, replace, treat or gradually eliminate all waste, including electronic waste. These figures would indicate that more than 80% of the companies analyzed report on the objectives related to improvements in efficiency in the use of resources and waste generation. In addition, more than 70% of companies indicate that they have proposed targets to reduce the overall ecosystem impact of the supply chain through joint and collaborative working. The remaining items show significant improvement in the number of companies required to start reporting.

Regarding the initiatives established by the companies regarding the use of resources and CE, the information related to the development of products that at the end of their useful life can generate positive effects on the environment (i.e., 100% biodegradable packaging, seed labels, etc.) is the most prominent among the reporting companies. In this sense, 61.4% of companies provide information in their reports. Next, 49.1% of companies present information related to the constant evaluation of the life cycle of materials in order to optimize their use.

Regarding resource input, companies mostly report on energy consumption from non-renewable sources in gigajoules and water consumption from natural sources in cubic meters, with an average frequency of 73% and 61.2%, respectively.

Regarding the output of resources, 58.3% of the companies highlight the information related to the total amount of non-hazardous waste they have produced. It should be noted that companies present information in a homogeneous manner for the remaining three items of this sub-score, relating to the total amount of waste produced, recycled or reused with a frequency of 47.6%, followed by the total amount of hazardous waste produced (47.5%) and the total amount of recycled plastic (47.4%).

Additionally, information related to proactive environmental investments or provisions to reduce future risks or increase future opportunities is reported by 33.2% of the European companies analyzed.

2.4 Dynamic Development of the Resources Use and Circular Economy Information Disclosure

Overview of the Disclosure of the Total Score: Year and Country

In order to provide more details to the general overview of the information reported by the companies, in panel a of Fig. 2.1, the interannual evolution of the disclosure of information on resource use and CE is observed, with 2017 being the year with the most information reported.

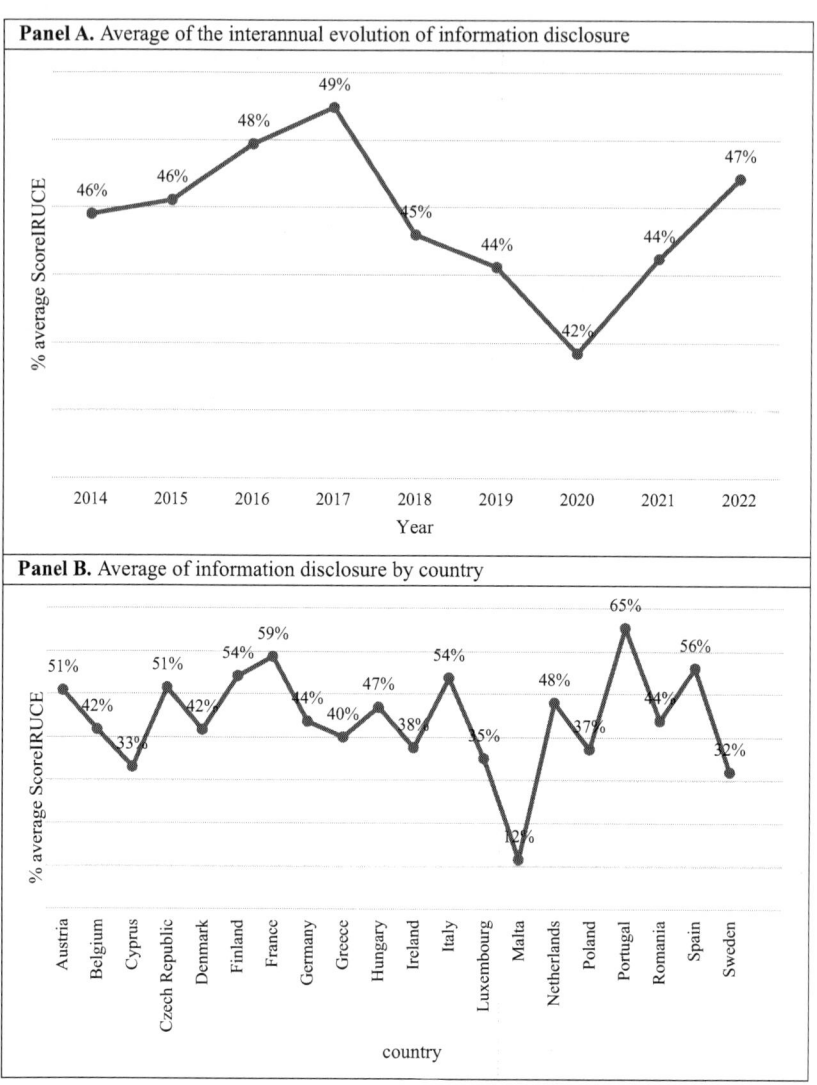

Fig. 2.1 Dynamic and geographic evolution of ScoreIRUCE: resources use and circular economy information disclosure by year and country. (**a**) Average of the interannual evolution of information disclosure. (**b**) Average of information disclosure by country. Source: Own elaboration

In this regard, companies report 49% of the disclosure items analyzed. However, in the three subsequent years, a slight decrease is observed in the data reported by companies, reaching a minimum of 42% average disclosure for 2020.

Subsequently, the companies increased the information disclosed, reaching a maximum average information of 47% of the items included in the ScoreIRUCE by 2022. In this way, during the nine years analyzed, the companies presented an average of 45% of progress in the disclosure of information required by current European regulations.

Panel b of Fig. 2.1 shows the average trend in the disclosure of information at the country level, standing out in first place, Portugal, whose companies report 65% of the items analyzed, followed by France with 59%, Spain with 56%, and Finland and Italy with 54%.

Overview of the Disclosure of Sub-scores: Year, Country and Industry

Panel a of Fig. 2.2 shows the temporal evolution of the four sub-scores and the FHP item described in Sect. 2.2, with 2017 being the year in which companies reported the most information for all sub-scores. In this sense, the information contained in the RFLs sub-score related to resource output and/or recycling reached an aggregate disclosure average of 50.64% during the 9 years analyzed, being in turn the most disclosed by the European companies that make up the study sample. Likewise, the TDOs sub-score related to goal setting had an average of 50.54% disclosure for the entire period analyzed.

Additionally, the INITs and RCMs sub-scores, and the FHP item had a homogeneous level of reporting, without being higher than 36% or lower than 34% of the information disclosure items during the analysis period.

Panel b of Fig. 2.2 shows the average percentage of information reported for each sub-score. For the information contained in the TDOs sub-score, Portugal stands out by disclosing an average of 69.70% of the data. Regarding the INITs sub-score, France stands out, with an average report of 48.33%.

In relation to the sub score-RCMs, Italy is the European country with the highest level of disclosure, reporting 51.09% of items analyzed. For the RFLs sub-score, Portugal stands out, whose companies present 84.60% of the information items considered. Finally, Spain stands out in the report of the items that make up the FHP score related to risk and opportunity forecasting, with 62.71% of the information considered.

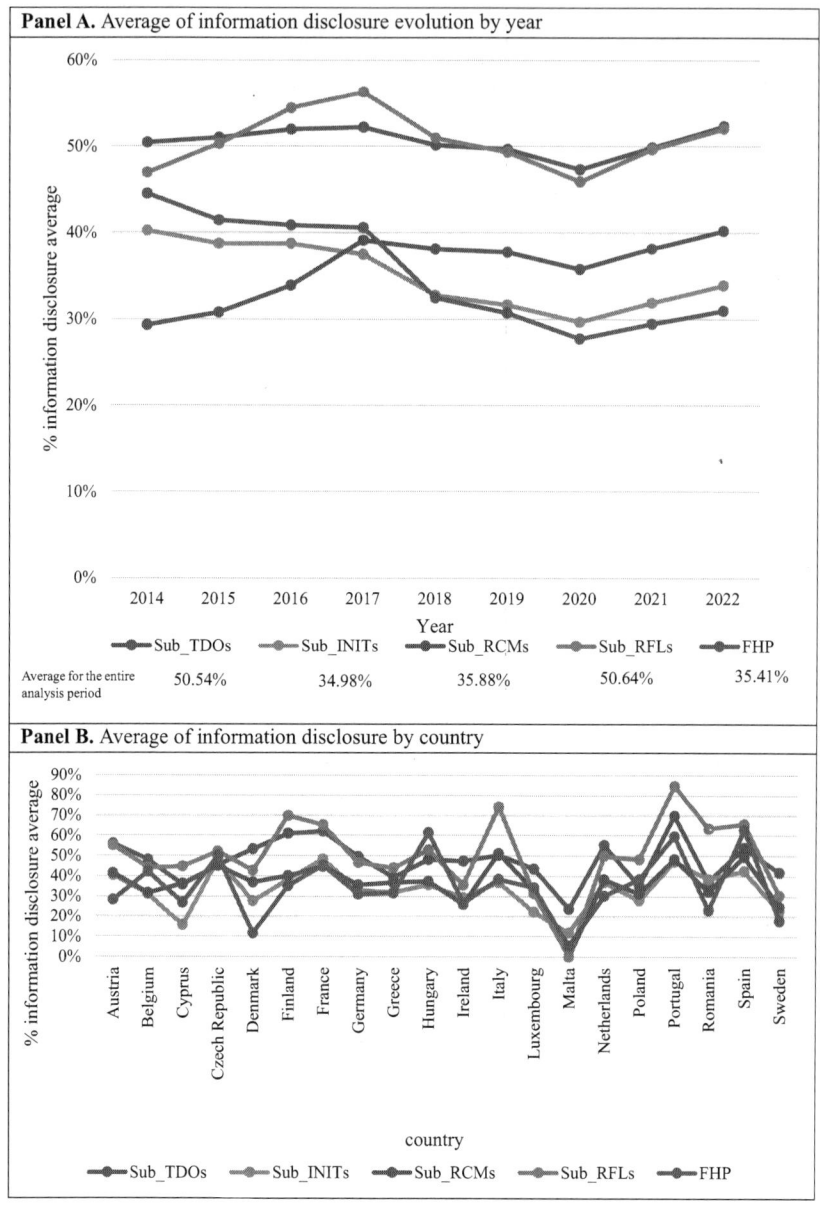

Fig. 2.2 Dynamic and geographic evolution of sub-scores of resources use and circular economy information disclosure by year and country. (**a**) Average of information disclosure evolution by year. (**b**) Average of information disclosure by country. Source: Own elaboration

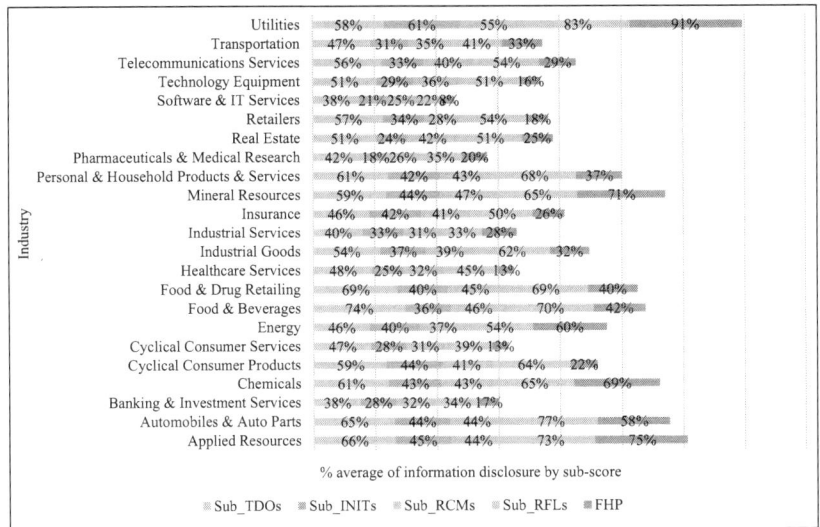

Fig. 2.3 Disclosure sub-scores for industry. Source: Own elaboration

In Fig. 2.3, you can see the average values taken by the sub-scores at the industry level. In this sense, the *Utilities industry* is the sector that reports the greatest volume of information for all sub-scores. Next, the *Applied resources* and *Automobiles and Auto parts* sectors stand out.

Vision of Information Dissemination at the Industrial Level

According to the European Environmental Agency (2024), companies generate pressure on the environment through the use of natural resources, the generation of waste or polluting emissions to ecosystems. These pressures are closely linked to the activities of each sector, identifying different levels of pressure or industrial risk. Following the approach of works such as that of Ali et al. (2024), the disclosure of corporate environmental information will be analyzed according to the level of sectoral risk. To do this, we have classified the companies in the sample into three risk levels: high, medium or low according to their economic activity (see Annex 1).

In this sense, panel a of Fig. 2.4 shows the percentage of companies in the sample classified as having a high, medium and low level of

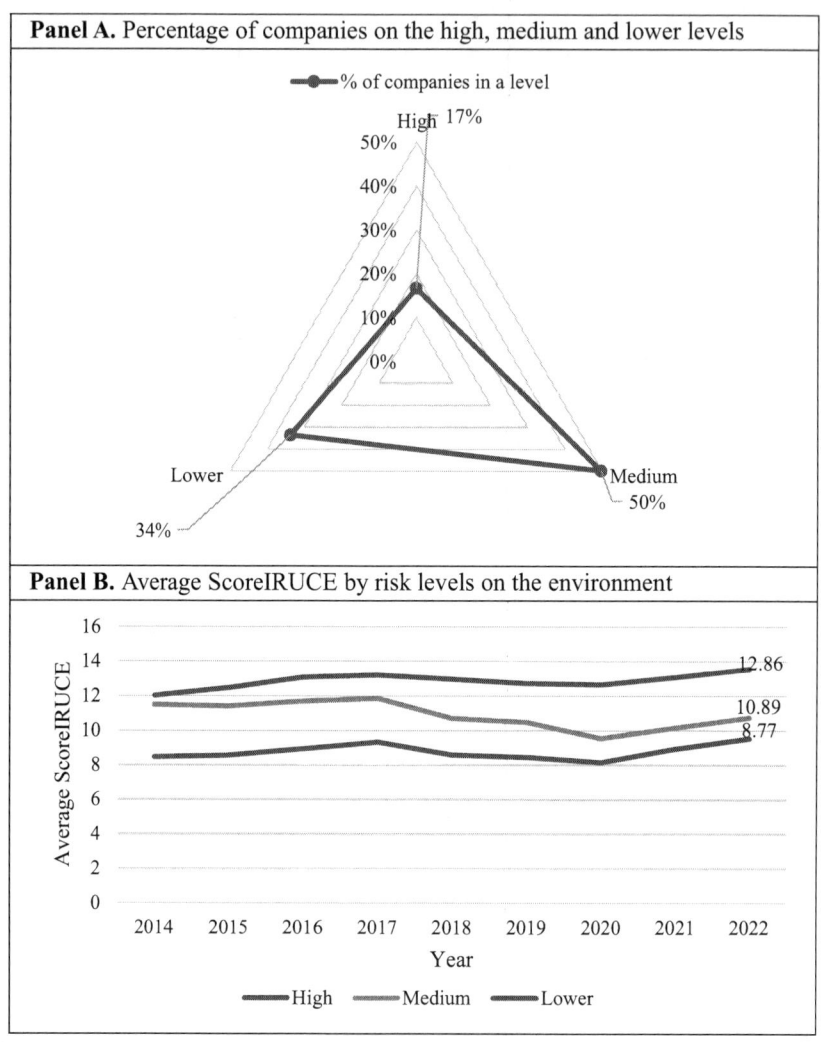

Fig. 2.4 Disclosure by industries risk levels on the environment. (**a**) Percentage of companies on the high, medium and lower levels. (**b**) Average ScoreIRUCE by risk levels on the environment

environmental risk, according to the industry to which they belong. Thus, 17% of the companies analyzed belong to an industry with a high risk of environmental impact, 50% are at a medium level of risk and 34% at a low level.

In this way, in panel b of Fig. 2.4, it can be seen that the companies with a high level of pressure on the environment are the ones that have disclosed the most information, reaching an aggregate average of 12.86 points out of 23 possible points, during the 9 years analyzed. Companies in industries with a medium level of environmental risk have obtained an average disclosure of 10.89 points and companies with low environmental risk have achieved an average score of 8.77 for the items included in the ScoreIRUCE. In this sense, companies with a high level of environmental risk report approximately 56% of the items analyzed. This frequency is reduced to 47% and 39% for the rest of the sectors, respectively.

2.5 Conclusions

The objective of this chapter is to analyze the information that European companies currently report on the use of resources and CE in accordance with the provisions of the Delegated Regulation that adopts the standards for the presentation of sustainability information. In this regard, ESRS E5 establishes the minimum information that companies must report as of its entry into force. This approach has allowed us to identify the gaps that companies face in order to align their sustainability reports with current demands.

In accordance with the purpose of this chapter, we observe that the information reported by the companies has fluctuated during the period analyzed. Although, as of 2020, companies present a sustained increase in the reporting of information on the use of resources and CE, facilitating dialogue with their stakeholders and the decisions that they can make in an economic environment that is committed to achieving circularity total for the year 2050.

In relation to the subcomponents considered in the analysis, we observe that European companies show greater progress in reporting information related to the output of resources and/or recycling of waste compared to the rest. Instead, business efforts should be directed toward improving information regarding CE and resource utilization initiatives, as well as companies' future risks and opportunities.

Additionally, at an industrial level, it is necessary to highlight the commitment of companies that operate in environmentally sensitive sectors with the disclosure of relevant information on the use of resources and CE given the level of risk associated with their activity. In this way, a correspondence was found between belonging to an industry with a high risk of environmental pressure and the volume of information that companies present to their stakeholders.

Finally, the analysis of the disclosure of the different items allowed us to observe that companies are making progress in reporting information, although in 2022 they reported 45% of the items that the Delegated Regulation proposes. In this regard, companies have room to increase the details and data related to efficiency goals for the use of water resources, the adoption of technologies for the treatment and/or purification of water that facilitates its recycling and the consequent organizational consumption of recycled water in processes and facilities, which would enrich, from a business perspective, the evaluation and monitoring of the capacity of natural water sources to meet human needs related to sustaining life and health and supporting sanitation and hygiene, and would ultimately serve as a proxy to evaluate the contribution business to maintaining adequate thresholds of natural processes such as freshwater change.FundingThis work was supported by Junta de Castilla y León y Fondo Europeo de Desarrollo Regional under Grant CLU-2019-03 Unidad de Excelencia "Gestión Económica para la Sostenibilidad" (GECOS). Servicio Público de Empleo Estatal (SEPE): Programa Investigo 2021.

Annex 1 Environment's Risk Classification for Industry

High-risk environmental impact
- Agriculture and livestock
- Mining, quarrying and coal
- Road transportation
- Oil and gas
- Food and beverages
- Textiles, accessories, footwear and jewelry
- Transport (road)
- Motor vehicles
- Energy production and utilities

(continued)

(continued)

Medium-risk environmental impact
- Forestry
- Chemicals
- Metal processing
- Building materials
- Construction and engineering
- Construction and furniture
- Defense
- Electronics
- Information technology
- Machinery and equipment
- Medical equipment
- Paper and wood products
- Biotechnology and pharmaceutical products
- Wholesale and retail trade
- Real estate
- Toys and sporting goods
- Tobacco
- Transport (other)

Water and sewerage services

Low-risk environmental impact
- Catering
- Capital markets
- Credit institutions
- Education
- Food and beverage services
- Gaming
- Health care and services
- Insurance
- Advertising and marketing
- Communications and media
- Professional services
- Recreation and leisure activities

Source: Own elaboration based on the information available at: https://www.efrag.org/

References

Ali, R., García-Sánchez, I.-M., Aibar-Guzmán, B., & Rehman, R. (2024). Is biodiversity disclosure emerging as a key topic on the agenda of institutional investors? *Business Strategy and the Environment, 33*(3), 2116–2142. https://doi.org/10.1002/bse.3587

Arvidsson, S., & Dumay, J. (2022). Corporate ESG reporting quantity, quality and performance: Where to now for environmental policy and practice?

Business Strategy and the Environment, 31(3), 1091–1110. https://doi.org/10.1002/bse.2937

Aureli, S., Gigli, S., Medei, R., & Supino, E. (2020). The value relevance of environmental, social, and governance disclosure: Evidence from Dow Jones Sustainability World Index listed companies. *Corporate Social Responsibility and Environmental Management, 27*(1), 43–52. https://doi.org/10.1002/csr.1772

Cubilla-Montilla, M. I., Galindo-Villardón, P., Nieto-Librero, A. B., Vicente Galindo, M. P., & García-Sánchez, I. M. (2020). What companies do not disclose about their environmental policies and what institutional pressures may do to respect them. *Corporate Social Responsibility and Environmental Management, 27*(3), 1181–1197. https://doi.org/10.1002/csr.1874

Desing, H., Brunner, D., Takacs, F., Nahrath, S., Frankenberger, K., & Hischier, R. (2020). A circular economy within the planetary boundaries: Towards a resource-based, systemic approach. *Resources, Conservation and Recycling, 155*, 104673. https://doi.org/10.1016/j.resconrec.2019.104673

DiMaggio, P. J., & Powell, W. W. (1983). The iron cage revisited: Institutional isomorphism and collective rationality in organizational fields. *American Sociological Review, 48*(2), 147–160. https://doi.org/10.2307/2095101

Enciso-Alfaro, S.-Y., & García-Sánchez, I.-M. (2024). Do boards care about planetary boundaries? A gender perspective on circular economy disclosures. Business Strategy and the Environment. https://doi.org/10.1002/bse.3700

European Commission. (2021, July 14). *The European Green Deal—European Commission.* https://commission.europa.eu/strategy-and-policy/priorities-2019-2024/european-green-deal_en

European Environmental Agency. (2024, January 25). *Industry.* https://www.eea.europa.eu/en/topics/in-depth/industry

European Parliament. (2014). Directive—2014/95—EN - NFRD - EUR-Lex. https://eur-lex.europa.eu/eli/dir/2014/95/oj

European Parliament. (2017). *Communication from the Commission—Guidelines on non-financial reporting (methodology for reporting non-financial information).*

European Parliament. (2019). *Communication from the Commission—Guidelines on non-financial reporting: Supplement on reporting climate-related information.*

European Parliament. (2020, May 20). *EU biodiversity strategy for 2030 | EUR-Lex.* https://eur-lex.europa.eu/EN/legal-content/summary/eu-biodiversity-strategy-for-2030.html

European Parliament. (2022). *Directive (EU) 2022/2464 of the European Parliament and of the Council of 14 December 2022 amending Regulation (EU) No 537/2014, Directive 2004/109/EC, Directive 2006/43/EC and Directive 2013/34/EU, as regards corporate sustainability reporting (Text with EEA relevance).* 322. http://data.europa.eu/eli/dir/2022/2464/oj/eng

European Parliament. (2023, July 31). *Commission Delegated Regulation (EU) 2023/2772 of 31 July 2023 supplementing Directive 2013/34/EU of* the European Parliament and of the Council as regards sustainability reporting standards. https://eur-lex.europa.eu/legal-content/EN/TXT/PDF/?uri=OJ:L_202302772

European Parliament. (2024, May 17). *How the EU wants to achieve a circular economy by 2050*. Topics | European Parliament. https://www.europarl.europa.eu/topics/en/article/20210128STO96607/how-the-eu-wants-to-achieve-a-circular-economy-by-2050

García-Sánchez, I.-M., Ortiz-Martínez, E., Marín-Hernández, S., & Aibar-Guzmán, B. (2023). How does the European Green Deal affect the disclosure of environmental information? *Corporate Social Responsibility and Environmental Management*, 30, 2766. https://doi.org/10.1002/csr.2514

García-Sánchez, I.-M., Sierra-García, L., & García-Benau, M.-A. (2022). How does the EU non-financial directive affect the assurance market? *Business Ethics, the Environment & Responsibility*, 31(3), 823–845. https://doi.org/10.1111/beer.12428

Geissdoerfer, M., Savaget, P., Bocken, N. M. P., & Hultink, E. J. (2017). The circular economy—A new sustainability paradigm? *Journal of Cleaner Production*, 143, 757–768. https://doi.org/10.1016/j.jclepro.2016.12.048

Kirchherr, J., Yang, N.-H. N., Schulze-Spüntrup, F., Heerink, M. J., & Hartley, K. (2023). Conceptualizing the circular economy (revisited): An analysis of 221 definitions. *Resources, Conservation and Recycling*, 194, 107001. https://doi.org/10.1016/j.resconrec.2023.107001

Korhonen, J., Honkasalo, A., & Seppälä, J. (2018). Circular economy: The concept and its limitations. *Ecological Economics*, 143, 37–46. https://doi.org/10.1016/j.ecolecon.2017.06.041

Liu, G., & Guo, L. (2023). How does mandatory environmental regulation affect corporate environmental information disclosure quality. *Finance Research Letters*, 54, 103702. https://doi.org/10.1016/j.frl.2023.103702

Martínez-Ferrero, J., & García-Sánchez, I.-M. (2017). Coercive, normative and mimetic isomorphism as determinants of the voluntary assurance of sustainability reports. *International Business Review*, 26(1), 102–118. https://doi.org/10.1016/j.ibusrev.2016.05.009

Mio, C., Fasan, M., Marcon, C., & Panfilo, S. (2020). The predictive ability of legitimacy and agency theory after the implementation of the EU directive on non-financial information. *Corporate Social Responsibility and Environmental Management*, 27(6), 2465–2476. https://doi.org/10.1002/csr.1968

Monteiro, A. P., Pereira, C., & Barbosa, F. M. (2021). Environmental disclosure on mandatory and voluntary reporting of Portuguese listed firms: The role of environmental certification, lucratively and corporate governance. *Meditari Accountancy Research*, 31(3), 524–553. https://doi.org/10.1108/MEDAR-09-2020-1001

Porkka, M., Virkki, V., Wang-Erlandsson, L., Gerten, D., Gleeson, T., Mohan, C., Fetzer, I., Jaramillo, F., Staal, A., te Wierik, S., Tobian, A., van der Ent, R.,

Döll, P., Flörke, M., Gosling, S. N., Hanasaki, N., Satoh, Y., Müller Schmied, H., Wanders, N., et al. (2024). Notable shifts beyond pre-industrial streamflow and soil moisture conditions transgress the planetary boundary for freshwater change. *Nature Water*, 2(3), 262–273. https://doi.org/10.1038/s44221-024-00208-7

Richardson, K., Steffen, W., Lucht, W., Bendtsen, J., Cornell, S. E., Donges, J. F., Drüke, M., Fetzer, I., Bala, G., von Bloh, W., Feulner, G., Fiedler, S., Gerten, D., Gleeson, T., Hofmann, M., Huiskamp, W., Kummu, M., Mohan, C., Nogués-Bravo, D., et al. (2023). Earth beyond six of nine planetary boundaries. *Science Advances*, 9(37), eadh2458. https://doi.org/10.1126/sciadv.adh2458

Rockström, J., Steffen, W., Noone, K., Persson, Å., Chapin, F. S., Lambin, E., Lenton, T. M., Scheffer, M., Folke, C., Schellnhuber, H. J., Nykvist, B., de Wit, C. A., Hughes, T., van der Leeuw, S., Rodhe, H., Sörlin, S., Snyder, P. K., Costanza, R., Svedin, U., et al. (2009). Planetary boundaries: Exploring the safe operating space for humanity. *Ecology and Society*, 14(2) https://www.jstor.org/stable/26268316

Steffen, W., Richardson, K., Rockström, J., Cornell, S. E., Fetzer, I., Bennett, E. M., Biggs, R., Carpenter, S. R., de Vries, W., de Wit, C. A., Folke, C., Gerten, D., Heinke, J., Mace, G. M., Persson, L. M., Ramanathan, V., Reyers, B., & Sörlin, S. (2015). Planetary boundaries: Guiding human development on a changing planet. *Science*, 347(6223), 1259855. https://doi.org/10.1126/science.1259855

United Nations, U. (2015). The Paris Agreement. United Nations. https://www.un.org/en/climatechange/paris-agreement

World Bank. (2024, July 1). *GDP ranking | Data Catalog*. https://datacatalog.worldbank.org/search/dataset/0038130

Wukich, J. J., Neuman, E. L., & Fogarty, T. J. (2023). Show me? Inspire me? Make me? An institutional theory exploration of social and environmental reporting practices. *Journal of Accounting & Organizational Change, ahead-of-print*. https://doi.org/10.1108/JAOC-01-2023-0013

Zampone, G., García-Sánchez, I.-M., & Sannino, G. (2022). Imitation is the sincerest form of institutionalization: Understanding the effects of imitation and competitive pressures on the reporting of sustainable development goals in an international context. *Business Strategy and the Environment*, 32, 4119. https://doi.org/10.1002/bse.3357

CHAPTER 3

Is It Just a Matter of Reputation? A Study on the Impact of ESG Controversies on Corporate Disclosure

Valentina Minutiello and Patrizia Tettamanzi

Abstract Due to the growing sensitivity toward ESG practices, academic attention on the subject has also grown. However, only a few studies have taken into consideration the influence that negative events can exert on the quality of corporate communication on environmental, social and governance issues.

The purpose of this chapter is, therefore, to fill the previous gap in the literature, studying the relationship between ESG disclosure and ESG controversies, from a legitimacy theory perspective. Findings confirm that companies may decide to improve the quality of their disclosure with the aim of restoring their legitimacy to operate in the market and mitigate the reputational impacts of adverse events.

Keyword ESG disclosure • Quality • ESG controversies • Legitimacy theory • Corporate social responsibility

---

V. Minutiello (✉) • P. Tettamanzi
Carlo Cattaneo University, LIUC, Castellanza, Italy
e-mail: vminutiello@liuc.it; ptettamanzi@liuc.it

© The Author(s), under exclusive license to Springer Nature Switzerland AG 2025
V. Minutiello (ed.), *The Development of Non-Financial Reporting*,
https://doi.org/10.1007/978-3-031-83181-2_3

3.1 Introduction

Today the relevance of firms' performance regarding environmental, social and governance (ESG) dimensions in the corporate value creation process is widely shared (Albarrak et al., 2019; Breuer et al., 2018; Bui et al., 2019; Richardson & Welker, 2001).

In fact, considering the impact of business decisions on the environment and society has become crucial. This impact can no longer be overlooked not only by governments and institutions but also by corporate managers. Consequently, ESG has become an increasingly relevant metric in the life of companies. In recent years, even investors have shown greater attention and sensitivity to ESG issues in defining their investment preferences (Lavin & Montecinos-Pearce, 2021). All these aspects have increased the pressure toward the communication of information on social, environmental and governance impacts and on the related company performance. For this reason, companies, especially large ones, are paying great attention to the issue of ESG disclosure to respond to the requests from investors, institutions and other stakeholders.

From these premises derives the birth of a thriving academic line of research. Some studies, for example, have investigated the different approaches of companies to sustainability, while others have focused on the different communication methods and the factors that determine improvements in the quality of corporate disclosure.

More recently, the need to clarify the extent to which this communication commitment derives solely from the search for legitimacy to operate by companies, thus simply resulting in opportunistic economic behavior and in an attempt to improve their reputation on the market (Schaltegger & Hörisch, 2015), has emerged. The absence of standardization of this type of disclosure, in fact, could favor its incorrect and not very transparent use (Chatterji et al., 2009; Escrig-Olmedo et al., 2014; Conca et al., 2021), and ample space is left to the discretion of managers, who, therefore, can decide whether and how to dispose of it and could use communication as a tool to overcome reputational problems or remedy previous corporate scandals (Aouadi & Marsat, 2018). These reflections are coherent with legitimacy theory and with the companies' need to gain legitimacy in carrying out their business.

This theme represents a significant gap in previous literature. ESG disclosure is a relatively recent research topic that, therefore, requires further studies and insights (Glass et al., 2016; Widyawati, 2020; Modugu, 2020). So far, there are only a few studies that have explored ESG as a whole (Siew et al., 2016; Yu & Van Luu, 2021), while most focus on just one of its three aspects (Aggarwal & Dow, 2011; Del Bosco & Misani, 2016; King & Lenox, 2000; Margolis & Walsh, 2003), just as few studies consider an international sample (Sanches Garcia et al., 2017; Yu & Van Luu, 2021). Such examples become even more scarce in numerical terms if we refer to ESG disclosure, rather than performance (Yu & Van Luu, 2021). Furthermore, few studies have investigated the issue of factors that can influence ESG disclosure and, for the most part, have considered internal variables of the company, such as corporate governance, ownership structure and firm size (e.g., Orazalin & Mahmood, 2019; Correa-Garcia et al., 2020; Songini et al., 2021), and not the external ones, such as the impact of previous scandals or negative news on the company which in turn are reflected in the disclosure (Aouadi & Marsat, 2018).

However, this topic is of fundamental importance, even more so since to date there is no unitary approach in terms of non-financial disclosure and its level of diffusion is still very varied from country to country and between different sectors (Ioannou & Serafeim, 2012; Reverte, 2009), thus making the comparability of the approaches and solutions adopted very complex. All these aspects make it necessary to further investigate ESG disclosure practices and their main drivers.

The aim of this study is to verify, in the light of the legitimacy theory, the potential influence of negative information on the companies' quality of ESG disclosure. Other possible determinants of the quality of the ESG disclosure were also taken into consideration, such as the country, the environmental sensitivity of the companies, the firm size, profitability and leverage, and some variables of corporate governance (board size, composition and gender diversity). An international sample of companies included in the S&P Global 1200 Index is analyzed over ten years (from 2010 to 2019). The data collection derives from the union of two different sources: Thomson Reuters Refinitiv for the ESG information and Bloomberg for all the other data. Our dependent variable is the ESG score which measures the quality of disclosure on ESG issues, and the independent variable is the ESG controversies score, which measures a company's exposure to environmental, social and governance controversies and negative events.

The structure of the chapter is as follows: firstly, we discussed the theoretical background and the hypothesis of our study; then we described the methodology; finally, the main results are presented; the last section provides the conclusions, the contributions and the limits of the study and proposes the future possible development of the research.

3.2 Background and Hypothesis

The ESG Disclosure and Its Theoretical Perspectives: The Role of Legitimacy Theory

ESG disclosure concerns the communication of information on three different types of issues: (1) environmental, in terms of impacts produced on the environment; (2) social, such as the health, safety and diversity of workers, the relationship with the community, the respect for human rights and so on; (3) governance, to guarantee a perfect balance between shareholders and stakeholder interests (Coetzee & van Staden, 2011; De Villiers et al., 2014; Holder-Webb et al., 2009).

More in general, this type of disclosure concerns the description of the adoption of an ESG strategy, or the application of policies by companies to achieve environmental and social performance objectives that reflect the needs of the community in which they operate and the requests of all stakeholders (Luo & Bhattacharya, 2006; Bresciani et al., 2016).

There may be several reasons behind the adoption of this type of strategy, also reflected in the related disclosure. The previous literature has expressed itself by interpreting such companies' behavior in the light of different theories.

According to the Resource-Based View (RBV), for example, undertaking environmental and social activities creates new competencies and skills in a company, thus increasing its competitive advantage over time (Hull & Rothenberg, 2008; Dressler & Paunović, 2019). The Agency Theory, on the other hand, claims the presence of agency costs, due to conflicting interests between the various stakeholders of the companies and information asymmetry between them (Fama & Jensen, 1983). The communication of ESG information helps to reduce these divergences, aligning the various interests and reducing agency costs.

The Agency Theory is also closely connected with the Stakeholder Theory: the performance of companies also depends on a series of relationships between different subjects, not just shareholders (Nekhili et al.,

2017) who, thanks to these relationships, produce sustainable wealth (Donaldson & Preston, 1995; Freeman, 2010; Mitchell et al., 1997). Non-financial disclosure, in this sense, helps companies achieve better performance and ensure their long-term survival (Gray et al., 2001), as it improves relations with their stakeholders and reduces the presence of information asymmetry (Gray et al., 1995; Ho & Taylor, 2007; Yakovleva & Vazquez-Brust, 2012). Furthermore, it also improves investor relations and, consequently, causes a reduction in the cost of capital (Ouyang et al., 2017) and produces a reduction in business risks and better financial performance (Cardoni et al., 2020; Cunha et al., 2020; Hoang et al., 2020; Jizi, 2017; Qureshi et al., 2020).

Another theory used to interpret the results related to corporate social responsibility (CSR) disclosure is the Signaling Theory (Dainelli et al., 2013; Rezaee, 2016; Uyar et al., 2020): performances are more likely to produce non-financial reports to communicate their positive results and thus distinguish themselves from their competitors. In this sense, disclosure is a sign of the achievement of a better performance that should influence future stakeholders' decisions.

Among these theories, the legitimacy theory is one of the most cited in the studies on CSR disclosure (Ortas et al., 2015) as it provides a robust interpretation of companies' behaviors (Modugu, 2020). It describes the relationship existing between companies and the community in which they operate in its entirety (without specifically referring to individual stakeholders) and establishes the desire on the part of companies to gain social acceptance to run their business (Deegan, 2002). Lindblom (1994, p. 2) provided the following definition of legitimacy: "a condition or status which exists when an entity's value system is congruent with the value system of the larger social system of which the entity is a part".

One of the most used tools by companies to establish their legitimacy is the adoption of non-financial reports, because they provide a series of social and environmental information in response to the pressures arising from the community (Mazzotta et al., 2020). Since, therefore, the companies need to demonstrate their legitimacy in carrying out business operations and given the growing attention and sensitivity of the community toward the issue of sustainability, their managers decide to use disclosure to emphasize companies' results, evaluating and reporting on their impacts on the environment and their contribution to the development of society (OECD, 2017; Majluf et al., 1998; Cornell, 2020).

In light of these considerations, the occurrence of unfavorable circumstances for the companies and the related negative effects represent a further strong motivation to increase and improve their disclosure to restore legitimacy and mitigate the reputational impacts of such negative events (Vourvachis et al., 2016). Several studies have confirmed this aspect, for example, with reference to the occurrence of safety incidents (Ouyang et al., 2017) or accidents in some industries (Knight & Pretty, 1996). In these cases, after the occurrence of the event, providing more information on safety policies allowed companies to reassure their stakeholders of the importance that they attribute to these circumstances and on the safeguard measures taken for the future (Coetzee & van Staden, 2011).

Communicating information on environmental, social and governance issues allows them to fully express their commitment in terms of sustainability and transparency, emphasizing their culture and values. The effects are the strengthening of the trust relationship with the stakeholders and the community in general and an improvement in the corporate image and reputation, with the restoration of legitimacy.

For all the above-mentioned different reasons, the number of companies that have adopted this type of disclosure has increased over time (Aureli et al., 2020; Cordazzo et al., 2020; Widyawati, 2020).

The Determinants of ESG Disclosure and the Importance of the ESG Controversies

Numerous authors have previously studied the behavior of companies toward the adoption of voluntary disclosure of non-financial information. Since their results often demonstrated a poor level of communication quality (Vormedal & Ruud, 2009; Boiral, 2013; Melloni, 2015; Melloni et al., 2016; Stacchezzini et al., 2016), they sought subsequent research to help companies identify those factors that could contribute to achieving an improvement in their reporting process.

Among the variables studied, a large space was given to the internal characteristics of companies, such as those relating to the corporate governance structure. In fact, the structure and composition of the Board of Directors (BoD) represent an important control tool capable of mitigating the presence of agency costs and improving companies' transparency. Several papers have confirmed that some corporate governance features can improve the quality of non-financial disclosure: for example, Cuadrado-Ballesteros et al. (2015), Albarrak et al. (2019) and Huafang and Jianguo

(2007) with reference to the proportion of independent directors; Tamimi and Sebastianelli (2017) concerning the board size and gender diversity; Barako et al. (2006) and Bravo and Reguera-Alvarado (2018) for the audit committee and so on. For example, Bravo and Reguera-Alvarado (2018) suggested a positive relationship between gender diversity in the audit committee and the quality of voluntary ESG reporting. Songini et al. (2021) found that integrated reporting (IR) quality is positively associated with the level of education of board members and negatively with the presence of women.

In addition to corporate governance, other internal variables analyzed in previous contributions are ownership structure (Huafang & Jianguo, 2007; Raimo et al., 2020; Correa-Garcia et al., 2020), firm size (Huafang & Jianguo, 2007; Barako et al., 2006; Frías-Aceituno et al., 2014; Ghani et al., 2018; Orazalin & Mahmood, 2019), and age (Correa-Garcia et al., 2020), profitability, leverage and industry (Barako et al., 2006; De Luca et al., 2020; Gonçalves et al., 2020; Lagasio & Cucari, 2019; Phan et al., 2020). For example, Raimo et al. (2020) highlighted a positive association between the quality of a sample of integrated reports and the presence of institutional ownership, while Barako et al. (2006) obtained a similar result in presence of foreign ownership.

It is possible to observe how previous studies have focused more on internal variables within the company, by considering only a limited number of external variables. Some authors have analyzed different country factors (e.g., Ioannou & Serafeim, 2012; Bui et al., 2019; Yu & Van Luu, 2021), such as the presence of different legal systems (Jensen & Berg, 2012) or national culture (Vitolla et al., 2019).

Among the external factors, the presence of negative CSR-related news stories is crucial for the definition of companies' behavior relating to the adoption of sustainability strategies and practices. However, to date, this topic has been not very deep in the literature.

Some authors have explored it concerning companies' value (Aouadi & Marsat, 2018). ESG controversies, in fact, decrease the firm value due to the loss of legitimacy and as a result of social control (e.g., Nirino et al., 2021). The presence of negative news undermines the relationship with the stakeholders, who show skepticism and low credibility toward the companies (Du et al., 2010; Maignan & Ralston, 2002; Godfrey et al., 2009; Yoon et al., 2006). Not only that but the loss of stakeholder trust also produces an increase in financial risk and the cost of capital (Lange & Washburn, 2012). Conversely, the more a company is perceived as virtuous, the more it will be appreciated by its stakeholders. Consequently, companies subject to controversies see their legitimacy completely compromised (Palazzo & Scherer, 2006).

The literature has also demonstrated that a positive CSR reputation can reduce the damage caused by negative news (Vanhamme & Grobben, 2009). This is even more true in presence of large companies, which are more visible and subject to social control (Aouadi & Marsat, 2018). In fact, several authors have shown how ESG practices improve the corporate image (Franceschelli et al., 2018; Santoro et al., 2019), their financial performance and their value creation process over time (Li et al., 2019).

However, this effect can also produce opportunistic behavior in companies, as explained by the legitimacy theory.

Companies may engage in CSR solely to restore their loss of reputation (Becker-Olsen et al., 2006; Livesey & Kearins, 2002). In this case, we speak of symbolic sustainable practices, i.e., actions undertaken by companies solely to show a positive image of themselves to the outside world and mitigate the impact of the occurrence of negative news (Kim et al., 2012). Such misconduct is also reflected in their communication process: therefore, companies will produce greater ESG disclosure solely to restore their legitimacy.

Based on the previous literature and in the light of the above reflections, we formulated the following hypothesis to be tested with the subsequent analysis:

HP: The presence of ESG controversies positively impacts the quality of ESG disclosure.

3.3 Method

Sample

The sample used to test the aforementioned hypothesis is made up of the 1200 international companies included in the S&P Global 1200 Index. Only companies for which the necessary data for the subsequent analysis were not available were excluded from this list. Therefore, the final sample consists of 1140 companies, located in 33 countries and belonging to 9 industries, for the period 2010–2019.

We have chosen this sample of companies as they are all large and, therefore, are more sensitive to external pressures from society due to their greater visibility. In addition, the sample is international and also includes various sectors: both of these aspects guarantee a better chance of generalizing the results obtained from the analysis.

Table 3.1 provides more details on the sample.

Table 3.1 Sample

Sample by industry			
Industry	%		
Basic materials	7.4%		
Communications	7.4%		
Consumer	33.3%		
Diversified	0.2%		
Energy	5.0%		
Financial	20.3%		
Industrial	14.3%		
Technology	7.0%		
Utilities	5.2%		
Total	100.0%		
Sample by geographic zone			
Zone	%	Zone	%
AS	0.2%	JN	12.3%
AU	3.9%	LX	0.4%
BE	0.8%	MX	0.9%
BZ	1.3%	NE	1.6%
CA	4.6%	NO	0.6%
CH	0.9%	NZ	0.1%
CL	0.9%	PE	0.2%
CO	0.2%	PO	0.2%
DE	1.1%	SI	0.3%
FI	0.8%	SK	1.1%
FR	4.3%	SP	1.3%
GB	7.8%	SW	2.3%
GE	3.6%	SZ	3.3%
HK	1.1%	TA	0.9%
IO	0.1%	U2	0.1%
IR	1.4%	US	40.6%
IT	1.1%	Total	100.0%

Variables and Methodology

For the data collection of the model, we referred to two different sources: Thomson Reuters Refinitiv for the ESG information and Bloomberg for the other data.

From the previous literature, it emerged the need to examine the environmental, social and governance aspects together and with a holistic approach because of their interconnection (Galbreath, 2013; Siew et al., 2016; Tamimi & Sebastianelli, 2017; Yu & Van Luu, 2021). For this reason, in this study, we used the Thomson Reuters Refinitiv ESG disclosure score to measure the CSR disclosure of the sample of companies. The choice of this score is motivated by the clarity in the calculation of the indicator and is consistent with other studies on the subject (e.g., Cheng et al., 2014; Durand & Jacqueminet, 2015; Dorfleitner et al., 2020). The score metrics are divided as follows: 68 on environmental issues (such as resource use, emissions and innovation), 30 on social issues (workforce, human rights, community and product responsibility) and 35 metrics on governance (shareholders and CSR strategy). The score range goes from 0 in the absence of ESG information up to 100 in the presence of an excellent level of ESG reporting.

The independent variable is the ESG controversies score, which measures a company's exposure to environmental, social and governance controversies and negative events. It is collected from Thomson Reuters Refinitiv, which defines the score as follows: "The ESG controversies score is calculated based on 23 ESG controversy topics and measures a company's exposure to environmental, social and governance controversies and negative events reflected in global media. During the year, if a scandal occurs, the company involved is penalized and this affects their overall ESGC scores and grading". Therefore, it contains the controversies concerning environmental, social and governance issues that occurred during the fiscal year (Aouadi & Marsat, 2018; Li et al., 2019).

This is a ranking percentile that captures the spread of unfavorable information disseminated by global media sources during a company's fiscal year. In particular, it includes 23 different ESG controversy topics, such as "controversies privacy" or "business ethics controversies". Therefore, this score takes into account the presence of negative events.

To complete the model we also included a list of control variables, collected by Bloomberg, typically used by previous studies on the quality of non-financial disclosure. They are described below.

Environmental sensitivity. We used a dummy variable (Envsen) assuming the value of 1 if the firm's activities produce a relevant impact on the environment (such as in the case of manufacturing and other industrial firms) and 0 otherwise.

Region. We assigned the value of 1 for Anglosaxon countries, 2 for Europe and 3 for other countries.

Firm size. We used the natural logarithm of the total assets (LnAssets) for each year considered.

Leverage. The variable leverage (Lev) was obtained from the net financial debt scaled by equity for each year considered.

Profitability. We used the return on equity (ROE) as an indicator of firms' profitability for every year.

BODsize. We chose the total number of Board of Directors (BoD) members.

BODGen. This control variable is expressed by the % of female directors within the BoD.

Nonex. It represents the number of non-executive directors within the BoD.

Since our dataset has a highly balanced panel structure, we opted to perform a panel analysis. Furthermore, thanks to the Breusch-Pagan LM test for random effects, we obtained further confirmation of the need to use a panel model rather than a simple Ordinary Least Squares (OLS) regression.

Then, the first step is to choose between fixed effect (FE) and random effect (RE). The first model, in studying the relationships between the variables, does not take into account the individual characteristics unchanged over time. Conversely, the second model also takes into account the effect of individual time-invariant characteristics.

To define which model to adopt in our study, we run the Hausman test. Its result suggested the use of an FE model.

The FE model is expressed in the following equation:

$$ESGDs_{it} = \alpha + \beta 1\, ESGCs_{it} + \beta 2\, Envsen_{it} + \beta 3\, Region_{it} + \beta 4\, LnAssets_{it} + \beta 5\, Lev_{it} + \beta 6\, ROE_{it} + \beta 7\, BoDsize_{it} + \beta 8\, BoDGen_{it} + \beta 9\, Nonex_{it} + \varepsilon_{it}.$$

where α is the constant, β are the estimated parameters, ε is the residual term, i represents the firm dimension and t the time period that ranges from 2010 to 2019.

3.4 Findings

Descriptive Statistics and Correlation Matrix

Table 3.2 summarizes the main descriptive statistics. Table 3.3 reports the average value of the variables for each year.

As can be seen from the tables, on average the quality of ESG disclosure is quite low (it reaches a score of 58.19 out of a total of 100), but there is an improvement over the period (it goes from 53.82 in 2010 to 64.43 in 2019). As regards, instead, the ESG controversies score, its average score is quite high (82.50) and remains fairly stable over the years: this means that, in general, a large amount of negative news that has affected companies has not been recorded.

Table 3.2 Descriptive statistics

Variable	Obs	Mean	Std. Dev.	Min	Max
ESGDs	10,389	58.19	19.29	1.76	94.82
ESGCs	10,437	82.50	29.41	0.44	100.00
LnAssets	11,149	23.62	1.61	15.35	28.71
Lev	10,930	180.35	4349.37	0	446,669.10
ROE	10,781	16.77	28.96	−132.89	1059.74
BoDsize	9922	11.25	3.08	0	33.00
BoDGen	10,887	18.37	12.48	0	71.43
Nonex	9883	2.85	2.60	0	21
Envsen	11,400	0.39	0.49	0	1
Region	11,400	1.37	0.65	1	3

Table 3.3 Descriptive statistics by year

Year	ESGDs	ESGCs	LNAssets	Lev	ROE	BoDsize	BoDGen	Nonex	Envsen	Region
2010	53.82	79.00	23.28	114.35	16.34	11.28	11.96	8.07	0.39	1.37
2011	54.85	80.71	23.36	533.64	16.37	11.38	12.91	8.07	0.39	1.37
2012	55.29	81.80	23.45	159.43	14.61	11.44	13.99	1.58	0.39	1.37
2013	55.52	80.84	23.46	119.69	15.19	10.86	15.26	1.72	0.39	1.37
2014	55.87	81.96	23.55	138.11	15.97	11.23	16.91	1.90	0.39	1.37
2015	57.93	88.20	23.69	191.16	16.76	11.37	18.91	2.12	0.39	1.37
2016	60.25	85.51	23.77	157.85	16.46	11.29	20.58	2.31	0.39	1.37
2017	61.55	85.27	23.79	115.84	18.76	11.28	22.20	2.50	0.39	1.37
2018	63.18	82.38	23.85	165.16	19.28	11.31	23.95	2.69	0.39	1.37
2019	64.43	77.29	23.96	116.89	17.78	11.17	25.76	2.83	0.39	1.37
Total	58.19	82.50	23.62	180.35	16.77	11.25	18.37	2.85	0.39	1.37

Before proceeding with the development of the model, the results of the correlation analysis (Table 3.4) are presented below, which demonstrate the absence of multicollinearity problems.

Panel Model

Table 3.5 summarizes the results of the fixed effects model to test the hypothesis proposed in this study.

As can be seen from the table, two variables (EnvSen and Region) are excluded from the model. This is consistent with the type of panel analysis, since the fixed effects model removes the variables that do not change over time, such as the industry and country.

The results of the equation show that the ESGCs variable (coeff. = −0.0208; p-value = 0.000) has a statistically significant negative relationship with the ESGD score.

It means that it is possible to accept the hypothesis: a decrease in the ESGC score implies a greater number of negative events, and this situation pushes companies to produce a higher quality ESG disclosure to mitigate the impacts.

This observation is also confirmed in the literature (Vourvachis et al., 2016; Ouyang et al., 2017; Mazzotta et al., 2020) and finds an interpretative confirmation in the legitimacy theory. In fact, the spread of negative news undermines the reputation of companies and damages relations with their stakeholders, thus producing worse results in terms of overall performance. Companies that have suffered unfavorable events use non-financial disclosure to respond to the pressures arising from society and restore their legitimacy, improving their image and reputation on the market (Vourvachis et al., 2016; Mazzotta et al., 2020).

With reference to the control variables, the results show that the quality of ESG disclosure is positively influenced by the firm size, profitability and two variables concerning the corporate governance structure (the % of women on the Board and the number of non-executive directors).

These findings are confirmed by previous literature. In particular, large firms tend to produce better disclosure as they are subject to greater external pressures (Frías-Aceituno et al., 2014; Ghani et al., 2018; Orazalin & Mahmood, 2019). As regards profitability, however, its improvement pushes companies to produce a quality disclosure to report the positive results achieved to the market, thus following the assumptions of the Signaling Theory (Orazalin & Mahmood, 2019; Pavlopoulos et al., 2019).

Table 3.4 Pearson correlation coefficients (*** p < 0.01, ** p < 0.05, * p < 0.1)

	ESGDs	ESGCs	Envsen	Region	LnAssets	Lev	ROE	BoDsize	BoDGen	Nonex
ESGDs	1									
ESGCs	−0.2617***	1								
Envsen	0.0326***	−0.0286***	1							
Region	0.0220**	−0.0388***	−0.0312***	1						
LnAssets	0.0368***	−0.0321***	−0.0052	0.0372***	1					
Lev	0.0123	0.0052	0.0081	0.0091	−0.0013	1				
ROE	0.0142	0.0012	−0.0159*	−0.0192**	−0.1088***	0.1996***	1			
BoDsize	−0.0180*	−0.0018	−0.0320***	0.0118	0.4258***	0.0205**	−0.0230**	1		
BoDGen	0.0900***	−0.0167*	0.0016	0.0145	0.1532***	−0.0102	0.0892***	0.0540***	1	
Nonex	−0.0116	−0.0351***	−0.0111	0.0174*	0.1230***	0.0079	0.0380***	0.2445***	0.3276***	1

Table 3.5 Panel fixed effects regression

ESGDs	Coef.	Std. Err.	t	P>t	[95% Conf. Interval]	
ESGCs	−0.0207706	0.0046811	−4.44	0.000	−0.029947	−0.0115941
Envsen	0 (omitted)					
Region	0 (omitted)					
LnAssets	7.145037	0.3773226	18.94	0.000	6.405369	7.884705
Lev	0.0005407	0.0004259	1.27	0.204	−0.0002942	0.0013757
ROE	0.0108121	0.0052869	2.05	0.041	0.0004482	0.021176
BoDsize	0.0428712	0.0715473	0.60	0.549	−0.0973834	0.1831259
BoDGen	0.1880513	0.0155861	12.07	0.000	0.1574978	0.2186048
Nonex	0.1135182	0.0531133	2.14	0.033	0.0093997	0.2176366
_cons	−113.371	8.808126	−12.87	0.000	−130.6377	−96.1044
R2 within	0.1062					
R2 between	0.0005					
R2 overall	0.0004					
No. obs	8022					
No. firms	1140					

Finally, concerning the corporate governance structure, the results are confirmed by numerous studies (Mahmood & Orazalin, 2017; Gerwanski et al., 2019; García-Sánchez et al., 2019; Vitolla et al., 2020). More specifically, a higher number of non-executive directors improve the efficiency of the Board of Directors and can exert more pressure on management to increase disclosure (Wang & Hussainey, 2013; Fasan & Mio, 2017), while, in the case of women, a greater presence of female directors improves non-financial disclosure, due to their greater sensitivity toward social and environmental issues (Barako & Brown, 2008; Frias-Aceituno et al., 2012; Prado-Lorenzo & Garcia-Sanchez, 2010; Rao & Tilt, 2016).

3.5 Conclusions

The relevance of the three ESG dimensions (environmental, social and governance) in the life of companies for the creation of a lasting competitive advantage over time is now out of question. Added to this are the growing pressures exerted by society and stakeholders for a greater sensitivity of companies toward these issues (Lavin & Montecinos-Pearce, 2021), which are expressed also with the request for more information and insights on the corporate solutions adopted in this direction.

Consequently, there has been growing attention from the academic world on CSR issues, on their reporting methods and on the factors that can increase the quality of corporate communication. However, previous studies have often considered ESG disclosure not in its entirety but solely with reference to one of the three aspects (Aggarwal & Dow, 2011; Del Bosco & Misani, 2016; King & Lenox, 2000; Margolis & Walsh, 2003). Furthermore, an even smaller number of studies have devoted themselves to identifying the variables that impact disclosure, with a greater focus on internal variables rather than external ones (Orazalin & Mahmood, 2019; Correa-Garcia et al., 2020; Songini et al., 2021). The latter includes the dissemination of news connected to the occurrence of unfavorable events for companies. This circumstance, in fact, can produce significant results on corporate communication (Aouadi & Marsat, 2018), pushing companies to produce more disclosure and of higher quality to mitigate the negative effects of adverse events. The explanation of this opportunistic behavior is found within the legitimacy theory and in the need that companies have to earn legitimacy to operate in the market.

Given the aforementioned gap in the literature, this study represents a first attempt to clarify the potential role of the dissemination of negative information on the quality of ESG disclosure. For the analysis, we used an international sample of large companies included in the S&P Global 1200 Index for the period 2010–2019.

Findings showed that there is a statistically significant relationship between the ESG controversies score and the ESG disclosure score: consistently with the legitimacy theory, a decrease in the first score implies an increase in negative events of a company, a situation that causes an improvement in the second score and in the quality of the disclosure. Therefore, companies produce higher quality disclosure to mitigate the reputational impacts of adverse situations.

Furthermore, with reference to the control variables, the quality of the ESG disclosure is positively influenced by the firm size, profitability and by the % of female directors in the BoD and the number of non-executive members, thus confirming some previous studies (Frías-Aceituno et al., 2014; Mahmood & Orazalin, 2017; Ghani et al., 2018; Orazalin & Mahmood, 2019; Pavlopoulos et al., 2019; Gerwanski et al., 2019; García-Sánchez et al., 2019; Vitolla et al., 2020).

Among the main contributions of the study, it complements the previous literature on ESG disclosure, considering the relationship with a new external variable (the presence of ESG controversies). The results obtained are interesting for regulators and policymakers and suggest the need, in evaluating corporate disclosure, to take into consideration scores that are purified from the effects produced by unfavorable events and that mitigate, therefore, the opportunistic behavior that companies can put in place. This consideration could help stakeholders make more informed choices regarding the organizations they relate to. Second, the study has an international and longitudinal nature and, therefore, guarantees a good level of generalizability of the results.

However, the study also has some limitations. First, it only refers to large companies. In the future, it could be interesting to extend the analysis to small and medium-sized enterprises and their peculiarities to observe if different results are obtained. Furthermore, in this case, the ESG controversies are studied in their entirety, but it could be interesting to conduct a more detailed study that takes into consideration specific categories of the score, to identify which adverse events that produce a greater impact on the behavior of companies.

REFERENCES

Aggarwal, R., & Dow, S. (2011). Corporate governance and business strategies for climate change and environmental mitigation. *The European Journal of Finance, 18*(3–4), 1–20.

Albarrak, M. S., Elnahass, M., & Salama, A. (2019). The effect of carbon dissemination on cost of equity. *Business Strategy and the Environment, 28*, 1179–1198.

Aouadi, A., & Marsat, S. (2018). Do ESG controversies matter for firm value? Evidence from international data. *Journal of Business Ethics, 151*(4), 1027–1047.

Aureli, S., Del Baldo, M., Lombardi, R., & Nappo, F. (2020). Nonfinancial reporting regulation and challenges in sustainability disclosure and corporate governance practices. *Business Strategy and the Environment, 29*, 2392–2403. https://doi.org/10.1002/bse.2509

Barako, D. G., & Brown, A. M. (2008). Corporate social reporting and board representation: Evidence from the Kenyan banking sector. *Journal of Management & Governance, 12*(4), 309–324.

Barako, D. G., Hancock, P., & Izan, H. Y. (2006). Factors influencing voluntary corporate disclosure by Kenyan companies. *Corporate Governance: An International Review, 14*, 107–125.

Becker-Olsen, K. L., Cudmore, B. A., & Hill, R. P. (2006). The impact of perceived corporate social responsibility on consumer behaviour. *Journal of Business Research, 59*(1), 46–53.

Boiral, O. (2013). Sustainability reports as simulacra? A counter-account of A and A+ GRI reports. *Accounting, Auditing & Accountability Journal, 26*(7), 1036–1071.

Bravo, F., & Reguera-Alvarado, N. (2018). Sustainable development disclosure: Environmental, social, and governance reporting and gender diversity in the audit committee. *Business Strategy and the Environment, 28*, 418–429. https://doi.org/10.1002/bse.2258

Bresciani, S., Ferraris, A., Santoro, G., & Nilsen, H. R. (2016). Wine sector: companies' performance and green economy as a means of societal marketing. *Journal of Promotion Management, 22*(2), 251–267.

Breuer, W., Mueller, T., Rosenbach, D., & Salzmann, A. (2018). Corporate social responsibility, investor protection, and cost of equity: A cross-country comparison. *Journal of Banking and Finance, 96*, 34–55.

Bui, B., Moses, O., & Houqe, M. N. (2019). Carbon disclosure, emission intensity and cost of equity capital: Multi-country evidence. *Accounting and Finance*. https://doi.org/10.1111/acfi.12492

Cardoni, A., Kiseleva, E., & Lombardi, R. (2020). A sustainable governance model to prevent corporate corruption: Integrating anticorruption practices, corporate strategy and business processes. *Business Strategy and the Environment, 29*, 1173–1185. https://doi.org/10.1002/bse.2424

Chatterji, A. K., Levine, D. I., & Toffel, M. W. (2009). How well do social ratings actually measure corporate social responsibility? *Journal of Economics & Management Strategy, 18*(1), 125–169. https://doi.org/10.1111/j.1530-9134.2009.00210.x

Cheng, B., Ioannou, I., & Serafeim, G. (2014). Corporate social responsibility and access to finance. *Strategic Management Journal, 35*(1), 1–23.

Coetzee, C. M., & van Staden, C. J. (2011). Disclosure responses to mining accidents: South African evidence. *Accounting Forum, 35*, 232–246. https://doi.org/10.1016/j.accfor.2011.06.001

Conca, L., Manta, F., Morrone, D., & Toma, P. (2021). The impact of direct environmental, social, and governance reporting: Empirical evidence in European-listed companies in the agri-food sector. *Business Strategy and the Environment, 30*(2), 1080–1093.

Cordazzo, M., Bini, L., & Marzo, G. (2020). Does the EU directive on nonfinancial information influence the value relevance of ESG disclosure? Italian evidence. *Business Strategy and the Environment, 29*, 1–14. https://doi.org/10.1002/bse.2589

Correa-Garcia, J. A., Garcia-Benau, M. A., & Garcia-Meca, E. (2020). Corporate governance and its implications for sustainability reporting quality in Latin American business groups. *Journal of Cleaner Production, 260*, 121142.

Cornell, B. (2020). ESG investing: Conceptual issues. *Journal of Wealth Management, 23*, 61–69.

Cuadrado-Ballesteros, B., Rodrıguez-Ariza, L., & García-Sanchez, I. (2015). The role of independent directors at family firms in relation to corporate social responsibility disclosures. *International Business Review, 24*(5), 890–901.

Cunha, F. A. F., De Oliveira, E. M., Orsato, R. J., Klotzle, M. C., Cyrino Oliveira, F. L., & Caiado, R. G. G. (2020). Can sustainable investments outperform traditional benchmarks? Evidence from global stock markets. *Business Strategy and the Environment, 29*, 682–697. https://doi.org/10.1002/bse.2397

Dainelli, F., Bini, L., & Giunta, F. (2013). Signaling strategies in annual reports: Evidence from the disclosure of performance indicators. *Advances in Accounting, 29*(2), 267–277. https://doi.org/10.1016/j.adiac.2013.09.003

Deegan, C. (2002). The legitimising effect of social and environmental disclosures—A theoretical foundation. *Accounting, Auditing & Accountability Journal, 15*(3), 282–311. https://doi.org/10.1108/09513570210435852

Del Bosco, B., & Misani, N. (2016). The effects of cross-listing on the environmental, social, and governance performance of firms. *Journal of World Business, 51*(6), 977–990.

De Luca, F., Cardoni, A., Phan, H. T. P., & Kiseleva, E. (2020). Does structural capital affect SDGs risk-related disclosure quality? An empirical investigation of Italian large listed companies. *Sustainability, 12*(5), 1776, 1–20. https://doi.org/10.3390/su12051776

De Villiers, C., Low, M., & Samkin, G. (2014). The institutionalisation of mining company sustainability disclosures. *Journal of Cleaner Production, 84*, 51–58. https://doi.org/10.1016/j.jclepro.2014.01.089

Donaldson, T., & Preston, L. E. (1995). The stakeholder theory of the corporation: Concepts, evidence, and implications. *Academy of Management Review, 20*(1), 65–91. https://doi.org/10.5465/amr.1995.9503271992

Dorfleitner, G., Kreuzer, C., & Sparrer, C. (2020). ESG controversies and controversial ESG: About silent saints and small sinners. *Journal of Asset Management, 21*(5), 393–412.

Dressler, M., & Paunović, I. (2019). Towards a conceptual framework for sustainable business models in the food and beverage industry. *British Food Journal, 122*, 1421.

Du, S., Bhattacharya, C., & Sen, S. (2010). Maximizing business returns to corporate social responsibility (CSR): The role of CSR communication. *International Journal of Management Reviews, 12*(1), 8–19.

Durand, R., & Jacqueminet, A. (2015). Peer conformity, attention, and heterogeneous implementation of practices in MNEs. *Journal of International Business Studies, 46*(8), 917–937.

Escrig-Olmedo, E., Muñoz-Torres, M. J., Fernández-Izquierdo, M. A., & Rivera-Lirio, J. M. (2014). Lights and shadows on sustainability rating scoring. *Review of Managerial Science, 8*(4), 559–574. https://doi.org/10.1007/s11846-013-0118-0

Fama, E. F., & Jensen, M. C. (1983). Separation of ownership and control. *Journal of Law and Economics, 26*(3), 301–325.

Fasan, M., & Mio, C. (2017). Fostering stakeholder engagement: The role of materiality disclosure in integrated reporting. *Business Strategy and the Environment, 26*(3), 288–305.

Franceschelli, M. V., Santoro, G., & Candelo, E. (2018). Business model innovation for sustainability: A food start-up case study. *British Food Journal., 120*, 2483.

Freeman, R. E. (2010). *Strategic management: A stakeholder approach.* Cambridge University Press. https://doi.org/10.1017/CBO9781139192675

Frias-Aceituno, J. V., Rodriguez-Ariza, L., & Garcia-Sanchez, I. M. (2012). The role of the board in the dissemination of integrated corporate social reporting. *Corporate Social Responsibility and Environmental Management, 20*(4), 219–233.

Frías-Aceituno, J. V., Rodríguez-Ariza, L., & García-Sánchez, I. M. (2014). Explanatory factors of integrated sustainability and financial reporting. *Business Strategy and the Environment, 23*(1), 56–72.

Galbreath, J. (2013). ESG in focus: The Australian evidence. *Journal of Business Ethics, 118*(3), 529–541.

García-Sánchez, I. M., Suárez-Fernández, O., & Martínez-Ferrero, J. (2019). Female directors and impression management in sustainability reporting. *International Business Review, 28*(2), 359–374.

Gerwanski, J., Kordsachia, O., & Velte, P. (2019). Determinants of materiality disclosure quality in integrated reporting: Empirical evidence from an international setting. *Business Strategy and the Environment, 28*(5), 750–770.

Ghani, E. K., Jamal, J., Puspitasari, E., & Gunardi, A. (2018). Factors influencing integrated reporting practices among Malaysian public listed real property companies: A sustainable development effort. *International Journal of Managerial and Financial Accounting, 10*(2), 144–162.

Glass, C., Cook, A., & Ingersoll, A. R. (2016). Do women leaders promote sustainability? Analyzing the effect of corporate governance composition on environmental performance. *Business Strategy and the Environment, 25*(7), 495–511. https://doi.org/10.1002/bse.1879

Godfrey, P. C., Merrill, C. B., & Hansen, J. M. (2009). The relationship between corporate social responsibility and shareholder value: An empirical test of the risk management hypothesis. *Strategic Management Journal, 30*(4), 425–445.

Gonçalves, T., Gaio, C., & Costa, E. (2020). Committed vs opportunistic corporate and social responsibility reporting. *Journal of Business Research. Forthcoming, 115*, 417–427. https://doi.org/10.1016/j.jbusres.2020.01.008

Gray, R., Javad, M., Power, D. M., & Sinclair, C. D. (2001). Social and environmental disclosure and corporate characteristics: A research note and extension. *Journal of Business Finance & Accounting, 28*(3–4), 327–356. https://doi.org/10.1111/1468-5957.00376

Gray, R., Kouhy, R., & Lavers, S. (1995). Corporate social and environmental reporting A review of the literature and a longitudinal study of UK disclosure. *Accounting, Auditing & Accountability Journal, 8*, 47–77.

Ho, L. C. J., & Taylor, M. E. (2007). An empirical analysis of triple bottom-line reporting and its determinants: Evidence from the United States and Japan. *Journal of International Financial Management and Accounting, 18*, 123–150.

Hoang, T.-H., Przychodzen, W., Przychodzen, J., & Segbotangni, E. A. (2020). Does it pay to be green? A disaggregated analysis of U.S. firms with green patents. *Business Strategy and the Environment, 29*, 1331–1361. https://doi.org/10.1002/bse.2437

Holder-Webb, L., Cohen, J. R., Nath, L., & Wood, D. (2009). The supply of corporate social responsibility disclosures among U.S. firms. *Journal of Business Ethics, 84*, 497–527. https://doi.org/10.1007/s10551-008-9721-4

Huafang, X., & Jianguo, Y. (2007). Ownership structure, board composition and corporate voluntary disclosure: Evidence from listed companies in China. *Managerial Auditing Journal, 22*, 604–619.

Hull, C. E., & Rothenberg, S. (2008). Firm performance: The interactions of corporate social performance with innovation and industry differentiation. *Strategic Management Journal, 29*(7), 781–789.

Ioannou, I., & Serafeim, G. (2012). What drives corporate social performance? The role of national-level institutions. *Journal of International Business Studies, 43*(9), 834–864.

Jensen, J. C., & Berg, N. (2012). Determinants of traditional sustainability reporting versus integrated reporting. An institutionalist approach. *Business Strategy and the Environment, 21*, 299–316.

Jizi, M. (2017). The influence of board composition on sustainable development disclosure. *Business Strategy and the Environment, 26*, 640–655. https://doi.org/10.1002/bse.1943

Kim, Y., Park, M. S., & Wier, B. (2012). Is earnings quality associated with corporate social responsibility? *The Accounting Review, 87*(3), 761–796.

King, A., & Lenox, M. (2000). *Does it really pay to be green? Accounting for strategy selection on the relationship between environment and financial performance.* New York University.

Knight, R. F., & Pretty, D. J. (1996). *The impact of catastrophes on shareholder value.* Oxford Exec. Res. Briefings. Templeton Coll, Univ of Oxford.

Lagasio, V., & Cucari, N. (2019). Corporate governance and environmental social governance disclosure: A meta-analytical review. *Corporate Social Responsibility and Environmental Management, 26*(4), 701–711. https://doi.org/10.1002/csr.1716

Lange, D., & Washburn, N. T. (2012). Understanding attributions of corporate social irresponsibility. *Academy of Management Review, 37*(2), 300–326.

Lavin, J. F., & Montecinos-Pearce, A. A. (2021). ESG reporting: Empirical analysis of the influence of board heterogeneity from an emerging market. *Sustainability, 13*(6), 3090.

Li, J., Haider, Z. A., Jin, X., & Yuan, W. (2019). Corporate controversy, social responsibility and market performance: International evidence. *Journal of International Financial Markets, Institutions and Money, 60*, 1–18.

Lindblom, C. K. (1994). The implications of organizational legitimacy for corporate social performance and disclosure. In *Critical perspectives on accounting conference*, New York.

Livesey, S. M., & Kearins, K. (2002). Transparent and caring corporations?: A study of sustainability reports by the body shop and Royal Dutch/Shell. *Organization and Environment, 15*(3), 233–258.

Luo, X., & Bhattacharya, C. B. (2006). Corporate social responsibility, customer satisfaction, and market value. *Journal of Marketing, 70*(4), 1–18.

Maignan, I., & Ralston, D. A. (2002). Corporate social responsibility in Europe and the US: Insights from businesses' self-presentations. *Journal of International Business Studies, 33*(3), 497–514.

Majluf, S. N., Abarca, M. N., Rodriguez, M. D., & Fuentes, L. (1998). Governance and ownership structure in Chilean economic groups. *Abante, 1*, 111–139.

Mahmood, M., & Orazalin, N. (2017). Green governance and sustainability reporting in Kazakhstan's oil, gas, and mining sector: Evidence from a former USSR emerging economy. *Journal of Cleaner Production, 164*, 389–397.

Margolis, J., & Walsh, J. (2003). Misery loves companies: Rethinking social initiatives by business. *Administrative Science Quarterly, 48*(2), 268–305.

Mazzotta, R., Bronzetti, G., & Veltri, S. (2020). Are mandatory non-financial disclosures credible? Evidence from Italian listed companies. *Corporate Social Responsibility and Environmental Management, 27*(4), 1900–1913. https://doi.org/10.1002/csr.1935

Melloni, G. (2015). Intellectual capital disclosure in integrated reporting: An impression management analysis. *Journal of Intellectual Capital, 16*(3), 661–680.

Melloni, G., Stacchezzini, R., & Lai, A. (2016). The tone of business model disclosure: An impression management analysis of the integrated reports. *Journal of Management & Governance, 20*(2), 295–320.

Mitchell, R. K., Agle, B. R., & Wood, D. J. (1997). Toward a theory of stakeholder identification and salience: Defining the principle of who and what really counts. *Academy of Management Review, 22*(4), 853–886. https://doi.org/10.5465/amr.1997.9711022105

Modugu, K. P. (2020). Do corporate characteristics improve sustainability disclosure? Evidence from the UAE. *International Journal of Business Performance Management, 21*(1–2), 39–54.

Nekhili, M., Nagati, H., Chtioui, T., & Rebolledo, C. (2017). Corporate social responsibility disclosure and market value: Family versus nonfamily firms. *Journal of Business Research, 77*(41), 52.

Nirino, N., Santoro, G., Miglietta, N., & Quaglia, R. (2021). Corporate controversies and company's financial performance: Exploring the moderating role of ESG practices. *Technological Forecasting and Social Change, 162*, 120341.

OECD. (2017). *Investment governance and the integration of environmental, social and governance factors.* OECD.

Orazalin, N., & Mahmood, M. (2019). Determinants of GRI-based sustainability reporting: Evidence from an emerging economy. *Journal of Accounting in Emerging Economies, 10*(1), 140–164.

Ortas, E., Alvarez, I., Jaussaud, J., & Garayar, A. (2015). The impact of institutional and social context on corporate environmental, social and governance performance of companies committed to voluntary corporate social responsibility initiatives. *Journal of Cleaner Production, 108*, 673–684. https://doi.org/10.1016/j.jclepro.2015.06.089

Ouyang, Z., Wei, J., & Zhao, D. (2017). Stock market's reaction to self-disclosure of work safety accidents: An empirical study in China. *Quality and Quantity, 51*, 1611–1626. https://doi.org/10.1007/s11135-016-035

Palazzo, G., & Scherer, A. G. (2006). Corporate legitimacy as deliberation: A communicative framework. *Journal of Business Ethics, 66*(1), 71–88.

Pavlopoulos, A., Magnis, C., & Iatridis, G. E. (2019). Integrated reporting: An accounting disclosure tool for high quality financial reporting. *Research in International Business and Finance, 49*, 13–40.

Phan, H. T. P., De Luca, F., & Iaia, L. (2020). The "walk" towards the UN sustainable development goals: Does mandated "talk" through nonfinancial disclosure affect companies' financial performance? *Sustainability, 12*(5). https://doi.org/10.3390/su12062324

Prado-Lorenzo, J. M., & Garcia-Sanchez, I. M. (2010). The role of the board in disseminating relevant information on greenhouse gases. *Journal of Business Ethics, 97*, 391–424.

Qureshi, M. A., Kirkerud, S., Theresa, K., & Ahsan, T. (2020). The impact of sustainability (environmental, social, and governance) disclosure and board diversity on firm value: The moderating role of industry sensitivity. *Business Strategy and the Environment, 29*, 1199–1214. https://doi.org/10.1002/bse.2427

Raimo, N., Vitolla, F., Marrone, A., & Rubino, M. (2020). The role of ownership structure in integrated reporting policies. *Business Strategy and the Environment, 29*, 2238–2250.

Rao, K., & Tilt, C. (2016). Board composition and corporate social responsibility: The role of diversity, gender, strategy and decision making. *Journal of Business Ethics, 138*(2), 327–347.

Reverte, C. (2009). Determinants of corporate social responsibility disclosure ratings by Spanish listed firms. *Journal of Business Ethics, 88*(2), 351–366.

Rezaee, Z. (2016). Business sustainability research: A theoretical and integrated perspective. *Journal of Accounting Literature, 36,* 48–64. https://doi.org/10.1016/j.acclit.2016.05.003

Richardson, A. J., & Welker, M. (2001). Social disclosure, financial disclosure and the cost of equity. *Accounting, Organisations and Society, 26*(7–8), 597–616.

Sanches Garcia, A., Mendes-Da-Silva, W., & Orsato, R. J. (2017). Sensitive industries produce better ESG performance: Evidence from emerging markets. *Journal of Cleaner Production, 150,* 135–147.

Santoro, G., Bresciani, S., Bertoldi, B., & Liu, Y. (2019). Cause-related marketing, brand loyalty and corporate social responsibility. *International Marketing Review, 37,* 773.

Schaltegger, S., & Hörisch, J. (2015). Search of the dominant rationale in sustainability management: Legitimacy- or profit-seeking? *Journal of Business Ethics, Forthcoming,* 1–18.

Siew, R. Y. J., Balatbat, M. C. A., & Carmichael, D. G. (2016). The impact of ESG disclosures and institutional ownership on market information asymmetry. *Asia- Pacific Journal of Accounting & Economics, 23*(4), 432–448.

Songini, L., Pistoni, A., Tettamanzi, P., Fratini, F., & Minutiello, V. (2021). Integrated reporting quality and BoD characteristics: An empirical analysis. *Journal of Management and Governance,* 1–42.

Stacchezzini, R., Melloni, G., & Lai, A. (2016). Sustainability management and reporting: The role of integrated reporting for communicating corporate sustainability management. *Journal of Cleaner Production, 136, Part A,* 102–110.

Tamimi, N., & Sebastianelli, R. (2017). Transparency among S&P 500 companies: An analysis of ESG disclosure scores. *Management Decision, 55*(8), 1660–1680.

Uyar, A., Karaman, A. S., & Kilic, M. (2020). Is corporate social responsibility reporting a tool of signaling or greenwashing? Evidence from the worldwide logistics sector. *Journal of Cleaner Production, 119997.*

Vanhamme, J., & Grobben, B. (2009). "Too good to be true!". The effectiveness of CSR history in countering negative publicity. *Journal of Business Ethics, 85*(2), 273–283.

Vitolla, F., Raimo, N., & Rubino, M. (2020). Board characteristics and integrated reporting quality: An agency theory perspective. *Corporate Social Responsibility and Environmental Management, 27*(2), 1152–1163.

Vitolla, F., Raimo, N., Rubino, M., & Garzoni, A. (2019). The impact of national culture on integrated reporting quality. A stakeholder theory approach. *Business Strategy and the Environment, 28*(8), 1558–1571.

Vormedal, I., & Ruud, A. (2009). Sustainability reporting in Norway - An assessment of performance in the context of legal demands and socio-political drivers. *Business Strategy and the Environment, 18*(4), 207–222.

Vourvachis, P., Woodward, T., Woodward, D. G., & Patten, D. M. (2016). CSR disclosure in response to major airline accidents: A legitimacy-based exploration. *Sustainability Accounting, Management and Policy Journal, 7*, 26–43. https://doi.org/10.1108/SAMPJ-12-2014-0080

Wang, M., & Hussainey, K. (2013). Voluntary forward-looking statements driven by corporate governance and their value relevance. *Journal of Accounting and Public Policy, 32*(3), 26–49.

Widyawati, L. (2020). A systematic literature review of socially responsible investment and environmental social governance metrics. *Business Strategy and the Environment, 29*, 619–637. https://doi.org/10.1002/bse.2393

Yakovleva, N., & Vazquez-Brust, D. (2012). Stakeholder perspectives on CSR of mining MNCs in Argentina. *Journal of Business Ethics, 106*, 191–211. https://doi.org/10.1007/s10551-011-0989-4

Yoon, Y., Gurhan-Canli, Z., & Schwarz, N. (2006). The effect of corporate social responsibility (CSR) activities on companies with bad reputations. *Journal of Consumer Psychology, 16*(4), 377–390.

Yu, E. P. Y., & Van Luu, B. (2021). International variations in ESG disclosure–Do cross-listed companies care more? *International Review of Financial Analysis, 75*, 101731.

Conclusions

This volume consists of a collection of empirical contributions on the topic of non-financial reporting, that is addressed from different perspectives. The main topics considered concern Intellectual Capital Disclosure, Circular Economy, and ESG controversies.

The chapters included aim to consider the research area from different points of view and study it with different methodologies.

Each contribution includes an overview of the macro-categories of the theme and provides food for thought on possible future developments and new issues to address.

Below, the main reflections arising from each chapter are summarized.

Concerning the first theme ("The factors that influence the quality of non-financial reports") the volume focuses on some internal and external factors. The contribution shows how these variables can have an impact on the quality of non-financial reports produced. However, other studies in the research area consider other internal variables of companies such as the composition of the main corporate governance bodies. Likewise, it could be interesting to delve deeper into the role of other variables (internal and/or external), as well as to develop new methodologies for measuring the quality of the reports analyzed.

The second theme, instead, regards the different types of non-financial disclosure. The volume takes into consideration Circular Economy

© The Author(s), under exclusive license to Springer Nature Switzerland AG 2025
V. Minutiello (ed.), *The Development of Non-Financial Reporting*, https://doi.org/10.1007/978-3-031-83181-2

disclosure. An increase in the number of information produced on the Circular Economy in the European context is recorded, as well as an increase in the quality of the information. This trend, however, is typical especially of companies most subject to environmental pressure. It should be noted that this is only an example of non-financial reporting, but other types of reports and communications can be found in the literature. Furthermore, it is interesting to observe other application contexts. An example could be Circular Economy disclosure in the public sector or by Universities or other specific categories of organizations. Furthermore, the volume also addresses the issue of Intellectual Capital Disclosure. Intangibles, in fact, have always represented a crucial resource for the competitive advantage of companies. The chapter provides some possible ideas to improve research on this topic and identifies some gaps in previous contributions, for example, in the applied methodology.

The third theme ("The motivations behind the adoption of non-financial communication") addresses the use of specific theories developed in the literature to explain the behavior of companies toward non-financial reporting. In this case, the legitimacy theory is taken into consideration, according to which companies can decide to use non-financial reporting to improve or restore their reputation on the market and toward their stakeholders. In this sense, non-financial reports can be used to mitigate the impacts of negative news regarding companies. However, many other theories in literature can be applied to interpret the behavior of companies, such as agency theory, signaling theory or stakeholders theory. Furthermore, the study refers to large companies, but it could also be interesting to consider small and medium enterprises or some particular categories of companies, such as family firms or benefit corporations.

All these ideas can be the subject of new analyses by the academic world and confirm the complexity of the topic, which still today requires further study and reflection. The volume has tried to focus attention on this aspect and, for this reason, could make the following contributions. It could be of particular interest to practitioners, with the aim of increasing their sensitivity to the topic. However, it can also be interesting for academics to identify new research possibilities and explore the main methodologies used for the study of non-financial reports. Furthermore, a topic that will certainly be studied in the coming years concerns the role of Artificial Intelligence as a tool to improve the quality of non-financial reporting and to support companies in the preparation of non-financial reports. Not only that, it will also be necessary to take into consideration new forms of communication of corporate information supported and introduced by the many digital tools available.

Finally, even if the approach is extremely practical and academic, it is the hope of the Authors that the volume can be the object of attention by a wider audience of subjects, which could include, for example, also students in economic and non-economic subjects, consumers, and, more generally, anyone who shows interest in the topic of the communication of non-financial information.

Index

A
Agency Theory, 23, 70, 94
Annual Report, 11, 14, 16, 17, 20–21, 25, 27, 28, 31, 32, 35

B
Bibliographic network analysis (BNA), 5, 6, 8–28

C
Circular Economy (CE), vi, 46–62, 93, 94
Citation network analysis (CNA), 5, 8–10, 18, 24, 33
Cluster, 25–27, 30, 34
Content analysis, 11, 14, 15, 17, 18, 25, 27, 31, 32, 34–36
Corporate Social Responsibility (CSR), 24, 71, 74, 76, 81
Corporate Sustainability Reporting Directive (CSRD), 50

D
Determinant, 11, 15, 16, 20, 21, 26, 27, 29, 30, 33–35, 69, 72–74
Disclosure, vi, 2–37, 46–62, 68–83, 93, 94

E
Environmental social governance (ESG), vi, 23, 68–83, 93
European Union (EU), 23, 46–62

F
Future research, 4, 17, 18, 30, 31, 33, 37

G
Global Citation Score (GCS) analysis, 4, 5, 18–22, 24, 30, 33
Global-Local Citation Score (GLCS), 5, 21–23, 30, 34
Global Reporting Initiative (GRI), 15, 21, 35, 51

H
Human Capital, 3, 7, 11, 15, 16, 23, 24, 35

I
Institutional theory, 49
Intangible, 3, 15, 18, 94
Intangible asset, 2, 3, 29, 35
Integrated Reporting (IR), v, 3, 4, 21, 23, 24, 29, 30, 34–36, 73
Intellectual Capital (IC), 3, 4, 7, 8, 11, 15–18, 21, 23–31, 34–36
Intellectual Capital Disclosure (ICD), vi, 2–37, 93, 94

K
Keywords, 4–7, 24–30, 33, 34, 37
Keywords Network Analysis (KNA), 6

L
Legitimacy theory, vi, 68–72, 74, 79, 82, 94
Literature review, vi, 5–8, 11, 17, 24–26, 28, 33, 37

M
Main Path, 5, 10–12, 18, 20, 21, 25, 30, 33, 35

O
Organizational Capital, 3, 7, 35

Q
Quality, v, vi, 3, 11, 15, 16, 20, 21, 23, 24, 30, 33, 34, 47, 50, 68, 69, 72–74, 76, 78, 79, 82, 93, 94

R
Relational Capital, 3, 7, 15, 31, 34, 36
Resource-Based View (RBV), 70

S
Signaling Theory, 71, 79, 94
Stakeholder Theory, 70
Structural Capital, 7, 11, 15, 26, 28, 34
Sustainability Reporting (SR), v, 35, 51
Systematic Literature Network Analysis (SLNA), vi, 2–37
Systematic literature review (SLR), 5–8, 33

9783031831805